Math Made Nice-n-Easy Books

In This Book:
- **Oblique Triangles**
- **Vectors**
- **Applications of Vectors**

"*MATH MADE NICE-n-EASY #6*" is one in a series of books designed to make the learning of math interesting and fun. For help with additional math topics, see the complete series of "*MATH MADE NICE-n-EASY*" titles.

Based on U.S. Government Teaching Materials

Research & Education Association
61 Ethel Road West
Piscataway, New Jersey 08854
Dr. M. Fogiel, Director

MATH MADE NICE-N-EASY BOOKS™
BOOK #6

Copyright © 2001 by Research & Education Association. This copyright does not apply to the information included from U.S. Government publications, which was edited by Research & Education Association.

Printed in the United States of America

Library of Congress Control Number 00-109062

International Standard Book Number 0-87891-205-3

MATH MADE NICE-N-EASY is a trademark of
Research & Education Association, Piscataway, New Jersey 08854

WHAT "MATH MADE NICE-N-EASY" WILL DO FOR YOU

The "Math Made Nice-n-Easy" series simplifies the learning and use of math and lets you see that math is actually interesting and fun. This series of books is for people who have found math scary, but who nevertheless need some understanding of math without having to deal with the complexities found in most math textbooks.

The "Math Made Nice-n-Easy" series of books is useful for students and everyone who needs to acquire a basic understanding of one or more math topics. For this purpose, the series is divided into a number of books which deal with math in an easy-to-follow sequence beginning with basic arithmetic, and extending through pre-algebra, algebra, and calculus. Each topic is described in a way that makes learning and understanding easy.

Almost everyone needs to know at least some math at work, or in a course of study.

For example, almost all college entrance tests and professional exams require solving math problems. Also, almost all occupations (waiters, sales clerks, office people) and all crafts (carpentry, plumbing, electrical) require some ability in math problem solving.

The "Math Made Nice-n-Easy" series helps the reader grasp quickly the fundamentals that are needed in using math. The reader is led by the hand, step-by-step, through the various concepts and how they are used.

By acquiring the ability to use math, the reader is encouraged to further his/her skills and to forget about any initial math fears.

The "Math Made Nice-n-Easy" series includes material originated by U.S. Government research and educational efforts. The research was aimed at devising tutoring and teaching methods for educating government personnel lacking a technical and/or mathematical background. Thanks for these efforts are due to the U.S. Bureau of Naval Personnel Training.

Dr. Max Fogiel
Program Director

Contents

Chapter 5
Oblique Triangles ... 1
   Aids to Calculation ... 2
   Law of Sines ... 2
   Law of Cosines ... 6
   Four Standard Cases ... 13
   Two Angles and One Side ... 14
   Three Sides ... 19
   Two Sides and the Included Angle ... 26
   Two Sides and an Opposite Angle ... 30
   Logarithmic Solutions ... 42
   Cases 1 and 4 ... 44
   Law of Tangents ... 51
   Case 2 ... 56
   Area Formula ... 56

Chapter 6
Vectors ... 63
   Definitions and Terms ... 63
   Scalars ... 63
   Vectors ... 64
   Symbols ... 64
   Combining Vectors ... 66
   Addition ... 66
   Subtraction ... 69
   Vector Solutions ... 70
   Graphical ... 70
   Analytical ... 73

Polar Coordinates 79
Multiplication 96
Division 98

Chapter 7
Applications of Vectors 105
    Statics 105
        Definitions and Terms 105
        Equilibrium 105
        Translation 106
        Rotation 106
        Translational Equilibrium 106
        First Condition 107
        Free Body Diagrams 115
        Rotational Equilibrium 122
        Second Condition 122
        Center of Gravity 127
    Additional Applications 128

Appendix
    Common Logarithms of Numbers 136
    Logarithms of Trigonometric Functions 138

Chapter 5
Oblique Triangles

The two previous chapters considered right angles and angles which could be calculated by using triangles which included right angles. This chapter considers oblique triangles which are, by definition, triangles containing no right angles.

Previously, a method for solving problems involving oblique triangles was introduced. This method employed the procedure of dividing the original triangle into two or more right triangles, and using the right triangles to solve the problem involved. It was also pointed out at that time athat there were direct methods of dealing with oblique triangles.

This chapter develops two methods of dealing directly with oblique triangles. These two methods, or laws, are developed in the first section of the chapter as aids to calculations. The chapter also contains example and practice problems for solving oblique triangles considered in four standard cases. In this chapter "solving a triangle" is defined as finding the three sides a, b, and c and the three angles A, B, and C of an oblique triangle, when some of these six parts are given.

Also included in this chapter are some problems using logarithms in solving oblique triangles (where another law is introduced) and problems concerning the area of a triangle which combine the area formula of plane geometry with the laws developed in this chapter.

Aids to Calculation

The aids to calculations, or aids in solving oblique triangles, developed in this chapter are two theorems, known as the law of sines and the law of cosines. This section is concerned with the development and proof of these laws; subsequent sections will be concerned with using them in calculations.

Law of Sines

The law of sines states that the lengths of the sides of any triangle are proportional to the sines of their opposite angles. If a triangle is constructed and labeled as shown in figure 5-1 the law of sines can be written

$$\frac{a}{\sin A} = \frac{b}{\sin B} = \frac{c}{\sin C}$$

PROOF: For the proof of the law of sines we redraw the triangle of figure 5-1 and drop a perpendicular from A to the opposite side as shown in figure 5-2 (A).

Reference to the triangle shown in figure 5-2 (A) shows that the perpendicular from A to the opposite side has divided the triangle into two right triangles and the trigonometric functions previously developed are used here. Considering these two right triangles we obtain

$$\sin B = \frac{h}{c} \text{ or } h = c \sin B$$

and,

$$\sin C = \frac{h}{b} \text{ or } h = b \sin C$$

Here we have two expressions for h which are equal to each other, so

$$c \sin B = b \sin C$$

or in another form

$$\frac{c}{\sin C} = \frac{b}{\sin B}$$

In figure 5-2 (B) the triangle is redrawn with a perpendicular from C to an extension of the opposite side. Considering the right triangle BCD thus formed, it is seen that

$$\sin B = \frac{h'}{a} \text{ or } h' = a \sin B$$

and in triangle ACD

$$\sin (180° - A) = \frac{h'}{b} \text{ or } h' = b \sin (180° - A)$$

From chapter 4 of this training course, recall that

$$\sin (180° - \theta) = \sin \theta$$

$h' = b \sin A$

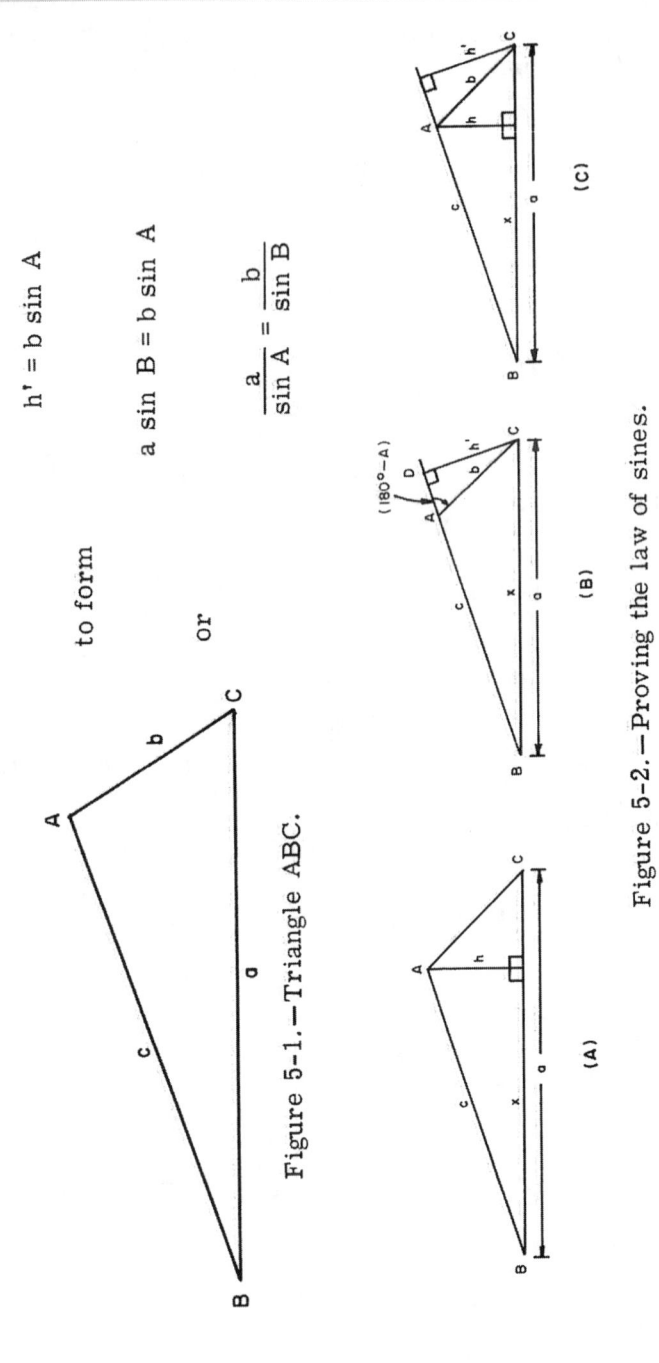

to form

$a \sin B = b \sin A$

or

$$\frac{a}{\sin A} = \frac{b}{\sin B}$$

Figure 5-1.—Triangle ABC.

Figure 5-2.—Proving the law of sines.

so

$$\sin(180° - A) = \sin A$$

then

$$\sin A = \frac{h'}{b} \text{ or } h' = b \sin A$$

Now equate

$$h' = a \sin B$$

and

Two separate triangles were used in proving the law of sines simply for clarity of explanation. If the two triangles are combined in one figure, as shown in figure 5-2 (C), it is seen that the two laws could be derived from this one illustration. Here it is obvious that the angle B which appears in both of the ratio pairs is the same angle; thus the ratios

$$\frac{c}{\sin C} = \frac{b}{\sin B}$$

and

$$\frac{a}{\sin A} = \frac{b}{\sin B}$$

can be combined to form the law of sines

Math Made Nice-n-Easy

$$\frac{a}{\sin A} = \frac{b}{\sin B} = \frac{c}{\sin C}$$

as previously stated.

Law of Cosines

The second of the laws to be developed in this section is the law of cosines which states: In any triangle the square of one side is equal to the sum of the squares of the other two sides minus twice the product of these two sides multiplied by the cosine of the angle between them. For the triangle in figure 5-3 (A), the law of cosines can be stated as

$$a^2 = b^2 + c^2 - 2bc \cos A$$

PROOF: Consider the triangle in figure 5-3 (B) with a perpendicular dropped from B to side b to form two right triangles. To prove

$$a^2 = b^2 + c^2 - 2bc \cos A$$

consider first the triangle ABD and note that

$$\cos A = \frac{x}{c} \qquad (1)$$

or

$$x = c \cos A \qquad (2)$$

and also

Law of Cosines

$$h^2 = c^2 - x^2 \qquad (3)$$

Substituting in (3) the value of x given in (2) results in

$$h^2 = c^2 - c^2 \cos^2 A \qquad (4)$$

In triangle BDC

$$h^2 = a^2 - (b - x)^2 \qquad (5)$$

which expands to

$$h^2 = a^2 - b^2 + 2bx - x^2 \qquad (6)$$

Substituting again for x the value given in (2) results in

$$h^2 = a^2 - b^2 + 2bc \cos A - c^2 \cos^2 A \qquad (7)$$

Equating the two values of h^2 in (4) and (7) gives

$$c^2 - c^2 \cos^2 A = a^2 - b^2 + 2bc \cos A - c^2 \cos^2 A \qquad (8)$$

Canceling like terms and rearranging,

$$c^2 - \cancel{c^2 \cos^2 A} = a^2 - b^2 + 2bc \cos A - \cancel{c^2 \cos^2 A}$$

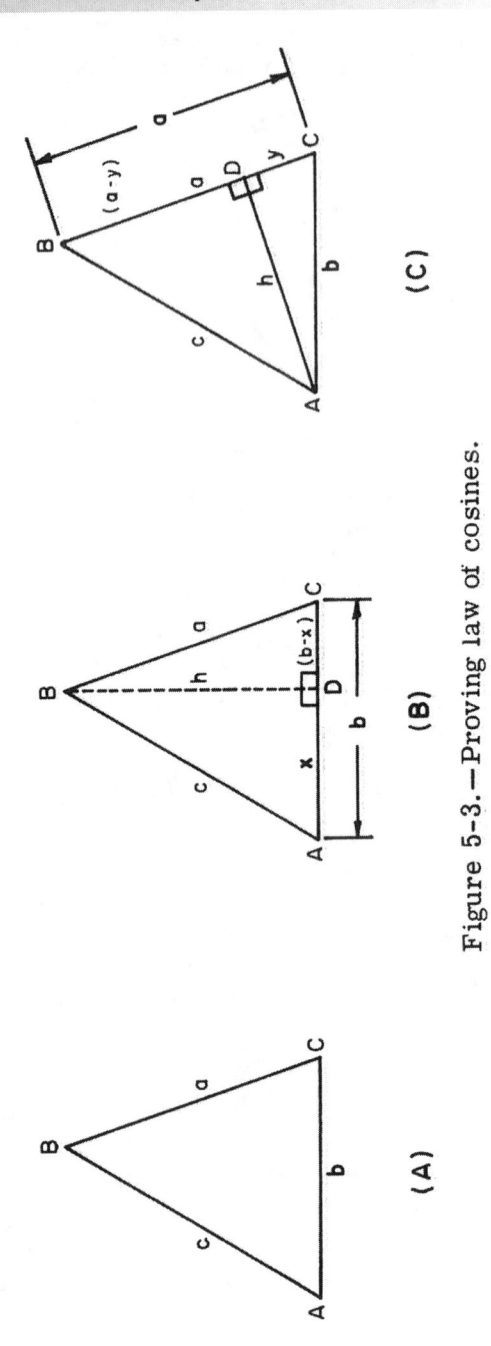

Figure 5-3.—Proving law of cosines.

Law of Cosines

$$-a^2 = -b^2 - c^2 + 2bc \cos A$$

$$a^2 = b^2 + c^2 - 2bc \cos A \qquad (9)$$

The proof is thus complete.

To prove another form of the law of cosines

$$c^2 = a^2 + b^2 - 2ac \cos C$$

refer to figure 5-3 (C) and note that in the right triangle CAD

$$\cos C = \frac{y}{b}$$

or

$$y = b \cos C$$

and also

$$h^2 = b^2 - y^2$$

Substituting for the value of y gives

$$h^2 = b^2 - b^2 \cos^2 C$$

In triangle ABD

$$h^2 = c^2 - (a - y)^2$$

which expands to

$$h^2 = c^2 - a^2 + 2ay - y^2$$

With additional substitution

$$h^2 = c^2 - a^2 + 2ab \cos C - b^2 \cos^2 C$$

Equating the two h^2 values which are representative of the two right triangles results in

$$b^2 - b^2 \cos^2 C$$

$$= c^2 - a^2 + 2ab \cos C - b^2 \cos^2 C$$

Canceling and rearranging yields

$$c^2 = a^2 + b^2 - 2ab \cos C$$

completing the proof.

The same procedures can be applied to prove the remaining form of the law of cosines. In summary, the three forms of the law of cosines are

$$a^2 = b^2 + c^2 - 2bc \cos A$$

$$b^2 = a^2 + c^2 - 2ac \cos B$$

$$c^2 = a^2 + b^2 - 2ab \cos C$$

The law of sines and the law of cosines are used mainly to solve oblique triangles, as will be shown in the following sections. In addition, these laws also hold true for right triangles.

Law of Cosines

The trigonometric functions or other methods previously noted are normally more effective in dealing with right triangles; however, application of these laws can be used in an analysis of some trigonometric principles and identities.

EXAMPLE: Show that

$$c^2 = a^2 + b^2 - 2ab \cos C$$

holds true in the right triangle shown in figure 5-4.

SOLUTION: In the figure it is shown that $C = 90°$. Recall from the graph of the cosine in chapter 4 of this course (or from appendix III) that $\cos 90° = 0$. Therefore, the formula

$$c^2 = a^2 + b^2 - 2ab \cos C$$

can be reduced to

$$c^2 = a^2 + b^2 - 2ab \cos 90°$$

$$c^2 = a^2 + b^2 - (2ab)(0)$$

$$c^2 = a^2 + b^2$$

Reference to the figure shows that c is the hypotenuse of a right triangle and from the Pythagorean theorem

$$c^2 = a^2 + b^2$$

Math Made Nice-n-Easy

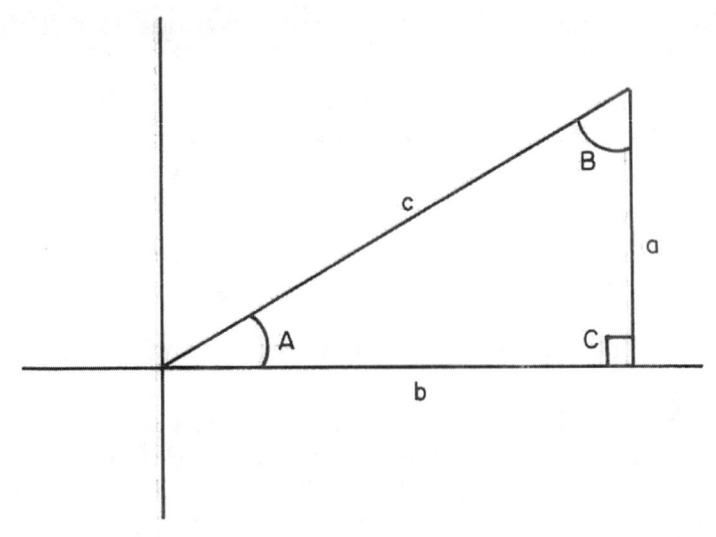

Figure 5-4.—Example problem, law of cosines.

Thus, the law of cosines would provide a proper solution to the right triangle.

In this problem the law of cosines reduces to the Pythagorean theorem. However, in working with oblique triangles remember that while the law of cosines applies to all triangles, the Pythagorean theorem can only be used when dealing with right triangles.

Practice Problems

1. Refer to figure 5-5 and prove that $b^2 = a^2 + c^2 - 2ac \cos B$.

2. Assume that the triangle in figure 5-5 is such that a=b=c=2. Transpose the formula in problem 1 and solve for cos B, then refer to the table of functions and verify that $B = 60°$.

Four Standard Cases

It was stated previously that the solution of a triangle consists of finding the six parts (sides a, b, and c; and the angles A, B, and C) when some of these values are known. If three of these parts are known, at least one of which is the length of a side, the remaining parts can normally be calculated by one of the methods discussed in the following paragraphs. For convenience, the methods for solving oblique triangles are developed by considering the triangles in four categories as follows:

1. Two of the angles and one of the sides are known.
2. The three sides are known.
3. Two of the sides and the angle between them are known.

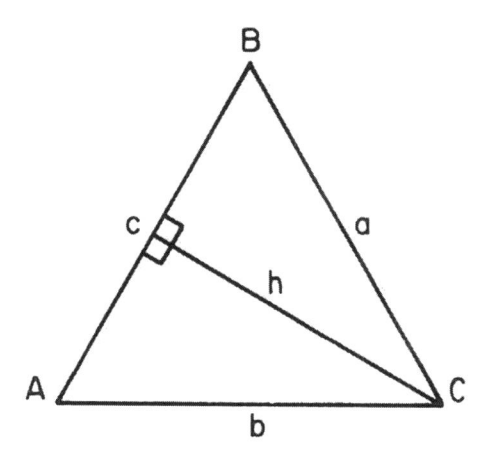

Figure 5-5.—Practice problem, law of cosines.

4. Two of the sides and an angle that is not between them are known.

The last situation described in the preceding list is known as the AMBIGUOUS CASE for, under certain conditions, two triangles which are not congruent can contain the same three known parts.

Recall from plane geometry that two triangles are congruent (having the same shape and size) if one of the following conditions is met:

1. Three sides of one triangle are equal to the corresponding sides of a second triangle.

2. Two sides and the included angle of one triangle are equal to the corresponding parts of a second triangle.

3. Two angles and a side of one triangle are equal to the corresponding parts of a second triangle.

It is seen here that the ambiguous case is the only one that does not parallel a plane geometry theorem for congruent triangles. The first and fourth (ambiguous) cases (or categories) of triangles will employ the law of sines in the solutions and cases 2 and 3 will be solved by using the law of cosines.

Two Angles and One Side

When two angles and a side are known, the remaining angle can be determined so easily that this case could be assumed to be one in which one side and all angles are known. (The third angle is equal to the difference between 180° and the sum of the known angles.) The law of sines is then used twice to find the length of the remaining sides. To find either of the unknown sides, select

Two Angles and One Side

the ratio pair which includes the ratio involving the unknown side and the one which considers the known side.

EXAMPLE: Using the law of sines, find the length of the lettered sides in the triangle in figure 5-6 (A).

SOLUTION: From the law of sines

$$\frac{c}{\sin C} = \frac{b}{\sin B}$$

$$\frac{5}{\sin 97.5°} = \frac{b}{\sin 30°}$$

Since

$$\sin \theta = \sin (180° - \theta),$$

$$\sin 97.5° = \sin 82.5° = 0.99144$$

Also

$$\sin 30° = 0.5000$$

so,

$$\frac{5}{0.99144} = \frac{b}{0.5000} \text{ or } \frac{2.50000}{0.99144} = b$$

or

$$b = 2.5216$$

Angle A is equal to

$$180° - B - C = 52.5°$$

Again from the law of sines,

$$\frac{c}{\sin C} = \frac{a}{\sin A}$$

$$\frac{5}{\sin 97.5°} = \frac{a}{\sin 52.5°}$$

$$\sin 52.5° = 0.79335$$

$$\frac{5}{0.99144} = \frac{a}{0.79335}$$

$$a = \frac{5(0.79335)}{0.99144}$$

$$a = 4.001$$

EXAMPLE: Figure 5-6 (B) shows a flagpole standing vertically on a hill which is inclined 15 degrees with the horizontal. A man climbing the hill notes that at one point his line of sight to the top of the pole makes an angle of 40° with the horizontal. At another point, 200 feet further up the hill, this angle has increased to 55°. How high is the flagpole? Solve using only the law of sines.

SOLUTION: First, define all the angles in the triangles OAB and OBD. In triangle OAB

$$\angle BAO = 40° - 15° = 25°$$

$$\angle OBA = 180° - (55° - 15°)$$

$$= 180° - 40° = 140°$$

$$\angle AOB = 180° - 140° - 25° = 15°$$

In triangle OBD

$$\angle DBO = 55° - 15° = 40°$$

$$\angle BDO = 90° + 15° = 105°$$

$$\angle BOD = 90° - 55° = 35°$$

These two triangles have OB as a common side. We can use the law of sines to find BO in triangle OAB and then apply the law again in triangle OBD to find the length of side OD which is the height of the flagpole. Thus,

$$\frac{AB}{\sin AOB} = \frac{OB}{\sin BAO}$$

$$\frac{200}{\sin 15°} = \frac{OB}{\sin 25°}$$

$$OB = \frac{200 \sin 25°}{\sin 15°} = 326.57 \text{ ft}$$

And in triangle OBD

$$\frac{OB}{\sin BDO} = \frac{OD}{\sin DBO}$$

$$\frac{326.57}{\sin 105°} = \frac{OD}{\sin 40°}$$

$$OD = \frac{326.57 \sin 40°}{\sin 105°}$$

Math Made Nice-n-Easy

$$\sin \theta = \sin(180° - \theta)$$

$$\sin 105° = \sin 75°$$

$$OD = \frac{326.57 \sin 40°}{\sin 75°}$$

$$OD = 217.3 \text{ ft}$$

Practice Problems

Refer to figure 5-7 in solving the following problems where the figures (A), (B), and (C) are to be used respectively with problems 1, 2, and 3, Use the law of sines in solving these problems.

1. Find a and b using the values given in the table of functions of special angles (chapter 4 of this course). Leave answers in radical form where applicable.

2. Find sides d and f to two decimal places.

3. Find the length of a to two decimal places.

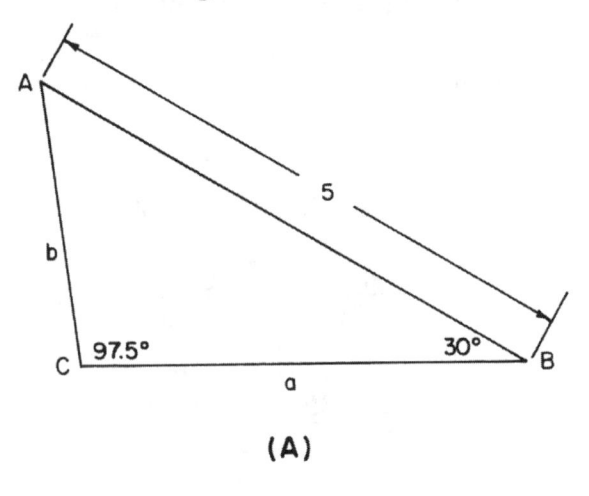

(A)

Three Sides

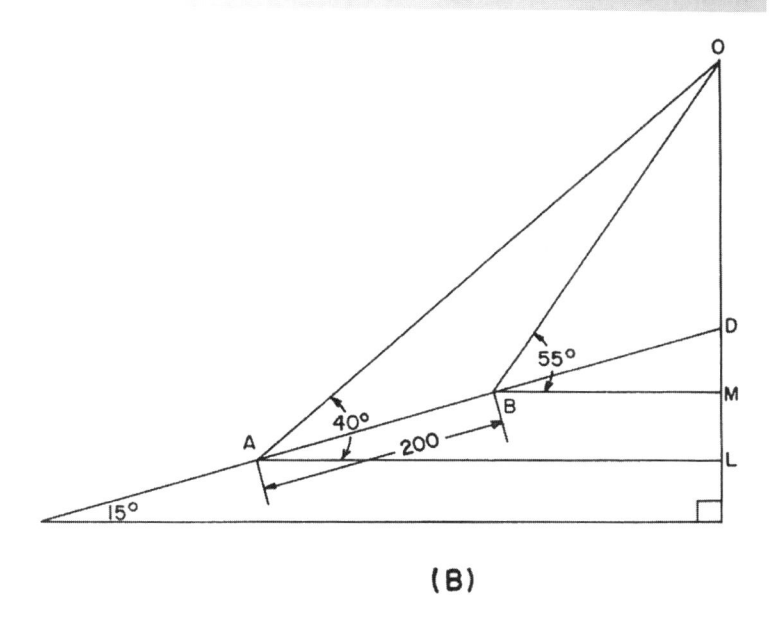

(B)

Figure 5-6.—Case 1, example problems.

Answers

1. a = 3
   b = 3√3
2. d = 6.07
   f = 3.96
3. 8.39

Three Sides

When the three sides of a triangle are given, the triangle can be solved by three successive applications of the law of cosines. Each application yields the value of one angle. The order of

Figure 5-7.—Case 1 practice problems.

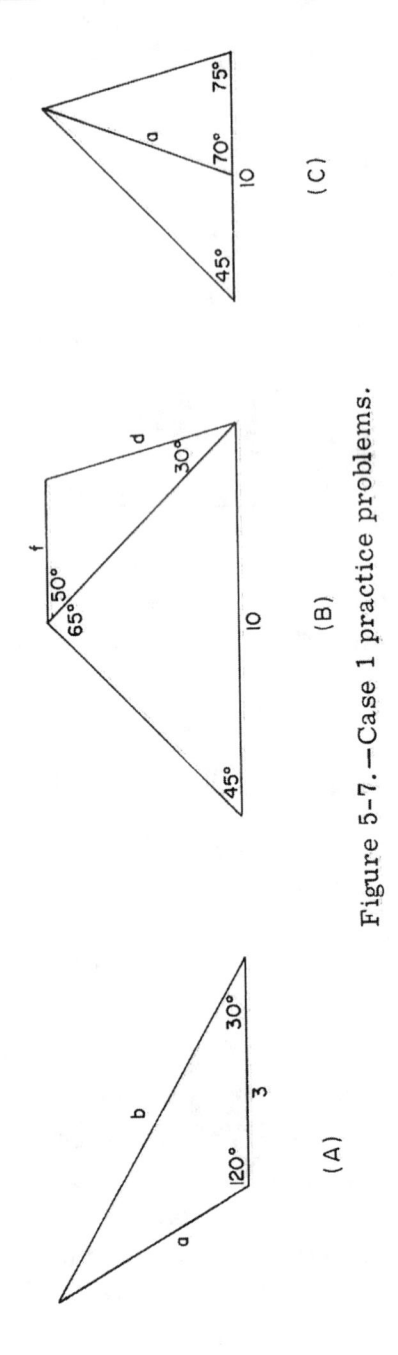

determining the angles is not important; any of the three angles may be determined first. A particular angle is found by using the form of the law of cosines in which the cosine of the angle in questions appears. When the three angles have been found, the solution is checked by verifying that $A + B + C \approx 180°$.

EXAMPLE: Solve the triangle ABC, given $a = 7$, $b = 13$, and $c = 14$. Determine the size of the angles to the nearest degree.

SOLUTION: To simplify the procedure solve the law of cosines algebraically for cos A.

$$a^2 = b^2 + c^2 - 2bc \cos A$$

$$2bc \cos A = b^2 + c^2 - a^2$$

$$\cos A = \frac{b^2 + c^2 - a^2}{2bc}$$

The remaining forms of the law can be solved in the same manner and the results are

$$\cos B = \frac{a^2 + c^2 - b^2}{2ac}$$

and

$$\cos C = \frac{a^2 + b^2 - c^2}{2ab}$$

Now in the given problem

$$\cos A = \frac{b^2 + c^2 - a^2}{2bc}$$

$$\cos A = \frac{13^2 + 14^2 - 7^2}{2 \times 13 \times 14}$$

$$\cos A = \frac{169 + 196 - 49}{364}$$

$$\cos A = \frac{316}{364}$$

$$\cos A = 0.86813$$

$$A = 30°$$

Then

$$\cos B = \frac{a^2 + c^2 - b^2}{2ac}$$

$$\cos B = \frac{49 + 196 - 169}{2 \times 7 \times 14}$$

$$\cos B = \frac{76}{196}$$

$$\cos B = 0.38776$$

$$B = 67°$$

and

$$\cos C = \frac{a^2 + b^2 - c^2}{2ab}$$

$$\cos C = \frac{49 + 169 - 196}{182}$$

$$\cos C = \frac{22}{182}$$

$$\cos C = 0.12088$$

$$C = 83°$$

Checking the calculations gives

$$A + B + C = 30° + 67° + 83° = 180°$$

It may appear that the best method to use in solving a triangle when three sides are given would be to calculate two angles and find the third angle by subtracting the sum of the two from 180°. While this method shortens the computation, it also destroys the check on the calculations, and is not recommended.

EXAMPLE: Solve the triangle ABC when $a = 8$, $b = 13$, and $c = 17$. Express the angles to the nearest degree.

SOLUTION:

$$\cos A = \frac{b^2 + c^2 - a^2}{2bc}$$

$$\cos A = \frac{169 + 289 - 64}{442}$$

$$\cos A = \frac{394}{442}$$

$$\cos A = 0.89140$$

$$A = 27°$$

Then

$$\cos B = \frac{64 + 289 - 169}{272}$$

$$\cos B = 0.67647$$

$$B = 47°$$

and finally

$$\cos C = \frac{64 + 169 - 289}{208}$$

$$\cos C = \frac{-56}{208}$$

$$\cos C = -0.26927$$

$$\cos C = -\cos 74°$$

$$-\cos C = \cos 74°$$

reference to reduction formulas gives

$$-\cos C = \cos(180° - C)$$

and

$$C = 106°$$

then, checking,

$$A + B + C = 27° + 47° + 106° = 180°$$

Practice Problems

The side lengths of triangle ABC are given in the following problems. Use the law of cosines to determine the sizes of the angles to the nearest degree.

1. $a = 3, b = 4, c = 5$
2. $a = 2, b = 3, c = 4$
3. $a = 7, b = 14, c = 11$

Answers

1. $A = 37°$
   $B = 53°$
   $C = 90°$
2. $A = 29°$
   $B = 47°$
   $C = 104°$
3. $A = 29°$
   $B = 100°$
   $C = 51°$

Two Sides and the Included Angle

Where two sides and the angle between them are given, the triangle is solved most easily by repeated use of the law of cosines. Using the given parts, solve first for the unknown side. Then, with three sides known, solve for the remaining angles in the same manner as in case two.

EXAMPLE: Using the law of cosines, solve the triangle ABC shown in figure 5-8, angle accuracy to the nearest degree.

SOLUTION: First find the unknown side.

$$a^2 = b^2 + c^2 - 2bc \cos A$$

$$a^2 = 7^2 + 5^2 - 2(5)(7) \cos 19°$$

$$a^2 = 49 + 25 - 70 \cos 19°$$

$$a^2 = 74 - 70(0.94552)$$

$$a^2 = 74 - 66.1864$$

$$a^2 = 7.8136$$

$$a = \sqrt{7.8136}$$

$$a = 2.795$$

To compute the angles, round the values given above to

$$\cos B = \frac{a^2 + c^2 - b^2}{2ac}$$

Two Sides and the Included Angle

$$\cos B = \frac{7.8 + 25 - 49}{2 \times 2.8 \times 5}$$

$$\cos B = -\frac{16.2}{28}$$

$$\cos B = -0.57857$$

$$B = 125°$$

and

$$\cos C = \frac{a^2 + b^2 - c^2}{2ab}$$

$$\cos C = \frac{7.8 + 49 - 25}{2 \times 2.8 \times 7} = \frac{31.8}{39.2}$$

$$\cos C = 0.81633$$

$$C = 35°$$

then, checking,

$$A + B + C = 19° + 125° + 35° = 179°$$

$$A + B + C \approx 180°$$

This is acceptable with the accuracy required here.

EXAMPLE: Two ships leave port at the same time; one (ship A) sailed on a course of 050° at a speed of 10 knots, the second (ship B) sails on a course of 110° at 12 knots. How far apart are the two ships at the end of 3 hours?

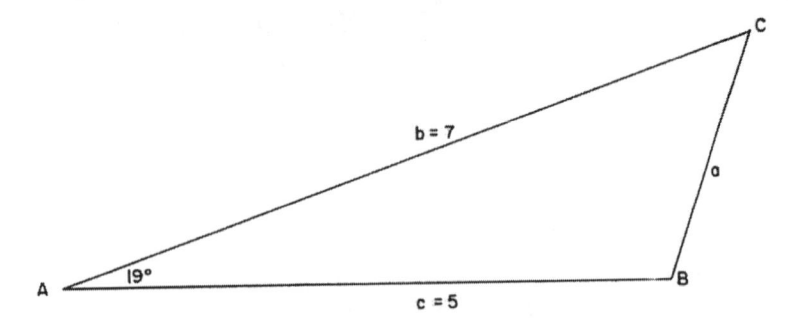

Figure 5-8.—Case 3, example problem.

SOLUTION: A good rule in any problem of this type is to first draw a picture to show the problem. In figure 5-9 a coordinate system oriented to compass headings is constructed, and the given information is plotted. From the figure it is seen that the desired answer is the distance AB opposite the angle labeled C, where C = 110° - 50° = 60°.

Using the law of cosines

$$c^2 = a^2 + b^2 - 2ab \cos C$$

$$c^2 = 36^2 + 30^2 - 2(30 \times 36) \cos 60°$$

$$c^2 = 1296 + 900 - \cancel{2}(1080) \times \frac{1}{\cancel{2}}$$

$$c^2 = 2196 - 180 = 1116$$

$$c = \sqrt{1116}$$

$$c = 33.5 \text{ approximately}$$

Two Sides and the Included Angle

The ships are approximately 33.5 miles part at the end of 3 hours.

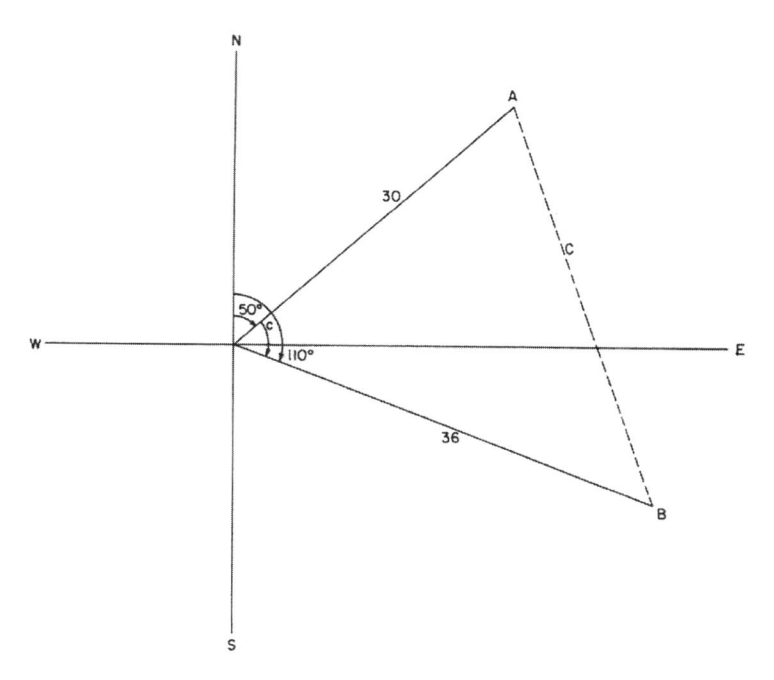

Figure 5-9.—Plot of ship's courses.

Practice Problems

Use the law of cosines to solve the triangles described below. Express angles to nearest degree and sides to two decimal places.

1. $a = 10$, $b = 7$, $C = 25°$
2. $b = 11$, $c = 17$, $A = 20°$
3. $a = 12$, $c = 26$, $B = 140°$

Answers

1. c = 4.69
   A = 116°
   B = 39°
2. a = 7.65
   B = 29°
   C = 131°
3. b = 36
   A = 12°
   C = 28°

Two Sides and an Opposite Angle

When two sides and a nonincluded angle are given, the triangle falls in the ambiguous category and one of the following cases will exist:
1. There is no solution.
2. There are two solutions.
3. There is one solution.

The category is called ambiguous for the given parts cannot always establish the shape and size of one triangle. There may be two triangles which are not congruent, but still contain the given parts. The ambiguity of this category can be seen if we assume that three parts (B, b, and c) are given, and we attempt to construct the triangle from this data. We consider the possibilities as follows:

1. If angle B is obtuse as in figure 5-10 (A), the side b must be larger than side c for a triangle to exist. In this case, b>c, there is only one triangle which exists and only one solution.

Two Sides and an Opposite Angle

2. If B is a right angle, as in figure 5-10 (B), side b must be larger than side c for a triangle to exist and there is only one triangle and one solution.

3. Figure 5-11 shows the situations which can exist if B is an acute angle. In (A) a figure is constructed with A < 90° and b < c; that is, a line drawn from vertex A is too short to reach the line BC. In this case, no triangle exists and there is no solution.

4. In figure 5-11 (B), the line from A (side b) is exactly the distance from A to the line BC. In this situation, only one triangle exists and it is a right triangle with one solution.

5. Figure 5-11 (C) shows a triangle where the line from A touches the line BC in two places. In this case the sides b and b' are longer than the side b in figure 5-11 (B), but still shorter than line c. In this category there are two triangles, BAC' and BAC, which contain the given parts. With the law of sines two solutions can be found for this possibility.

6. The last possibility considered is one shown in figure 5-11 (D). In this figure the line b is longer than c and again would touch the line BC in two places to form two triangles. However, only the triangle ABC is considered since the triangle ACC' does not include angle B as an interior angle. This situation is considered to have only one solution. If b = c, then c and c' coincide and there is only one triangle.

Certain relationships of angle, side, and function values can be found to determine in advance which of the possibilities previously listed exists for a given triangle. However, this knowledge is not required before the solution is attempted. It can be determined in the process of attempting a solution, as will be shown in the example problems, or a drawing can be made from the data given.

The triangles presented in this section have had value sizes only in relation to each other or in relation to 90°. Figure 5-12 points out the ambiguity which can exist when a triangle is described by giving two sides and an angle opposite one of these sides. This figure shows two triangles constructed with given data of $A = 30°$, $a = 4$, and $c = 6$. Solution of these two triangles will show that B, C, and b are not the same for the two, and this is a case where two solutions arise.

A good approach to solving triangles when two sides and an opposite angle (ambiguous category) are given is to first use the law of sines to find the unknown angle opposite a given side. Then the third angle can easily be determined and the law of sines can be used again, to compute the unknown side. In the following examples and practice problems desired accuracy is in degrees to the nearest minute and sides to two decimal places.

EXAMPLE: Solve the triangle (or triangles) ABC when $B = 45°$, $b = 3$, $c = 7$.

SOLUTION: First use the law of sines to find angle C

$$\frac{c}{\sin C} = \frac{b}{\sin B}$$

$$\sin C = \frac{c \sin B}{b}$$

Two Sides and an Opposite Angle

$$\sin C = \frac{7 \sin 45°}{3}$$

$$\sin C = \frac{7 \times 0.70711}{3}$$

$$\sin C = \frac{4.94977}{3}$$

$$\sin C = 1.64992$$

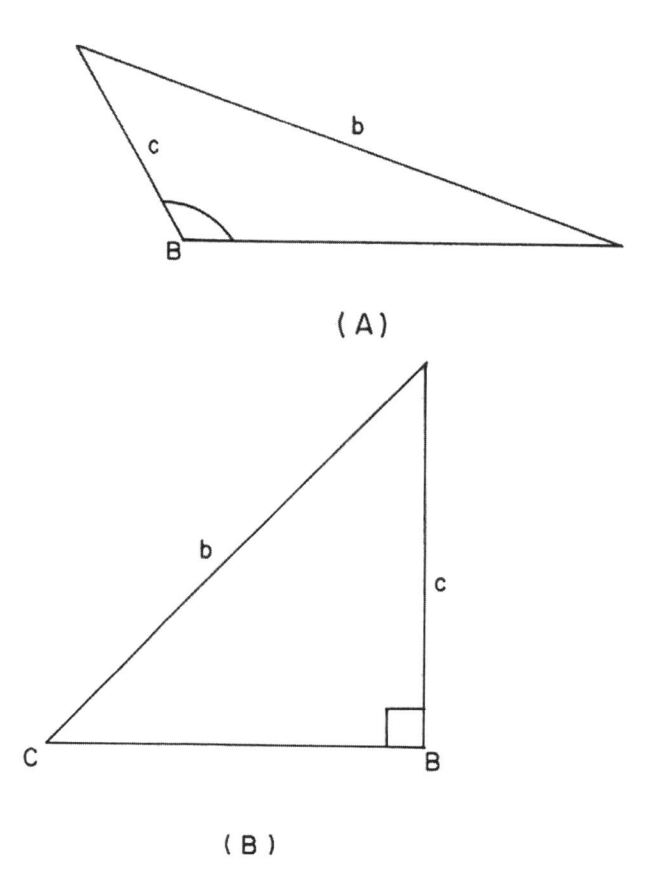

(A)

(B)

Figure 5-10.—Ambiguous case, $B \geq 90°$.

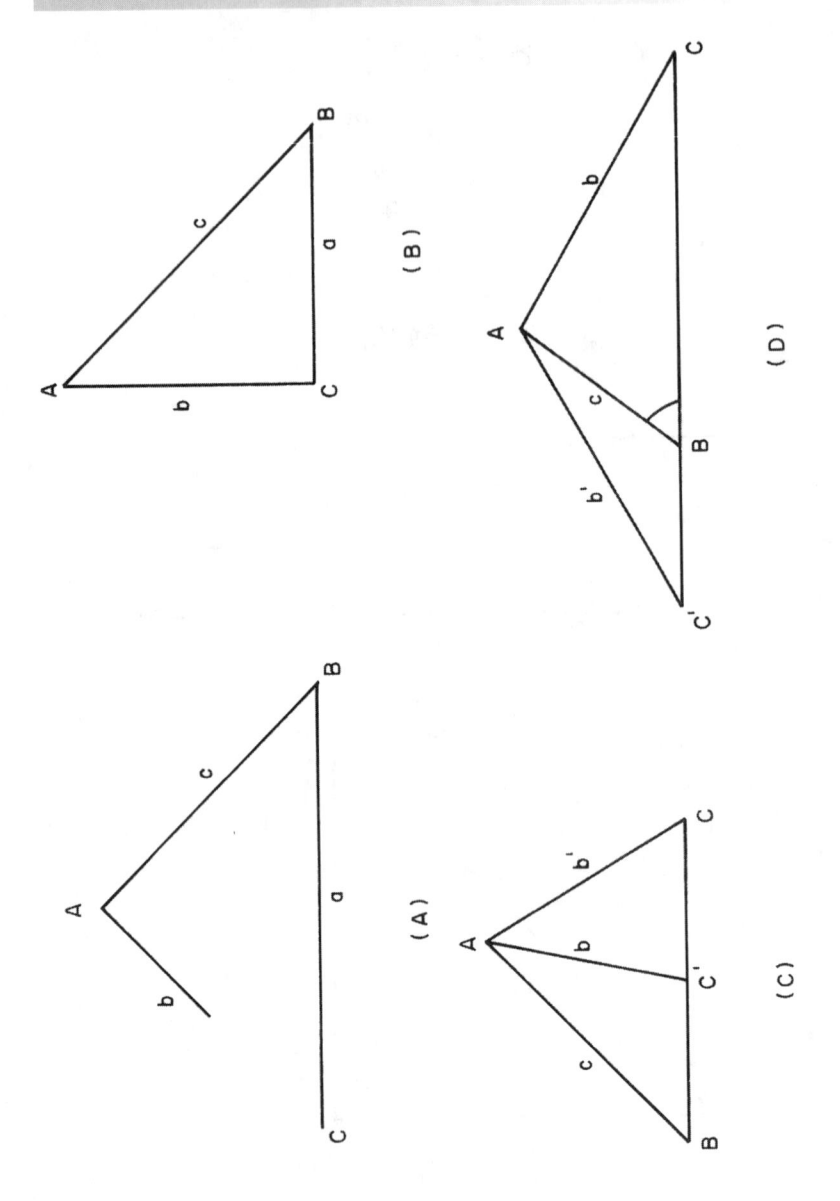

Figure 5-11.—Ambiguous case, B < 90°

Two Sides and an Opposite Angle

The calculations show that

$$\sin C > 1$$

However, reference to tables or a graph of the sine function shows that the sine is never greater than 1, so this is the case where no triangle or solution exists.

EXAMPLE: Solve the triangle (or triangles) ABC when $A = 22°$, $a = 5.4$, $c = 14$.

SOLUTION: Apply the law of sines to determine angle C.

$$\frac{c}{\sin C} = \frac{a}{\sin A}$$

$$\sin C = \frac{c \sin A}{a}$$

$$\sin C = \frac{14 \sin 22°}{5.4}$$

$$\sin C = \frac{14 \times 0.37461}{5.4}$$

$$\sin C = 0.97121$$

$$C = 76° 13'$$

Since the side opposite the known angle is smaller than the other given side, there are two angles to consider. Since

$$\sin(180° - C) = \sin C$$

the other angle (C') is 103° 47'.
Continue the solution considering two triangles, ABC and A'B'C'. In ABC;

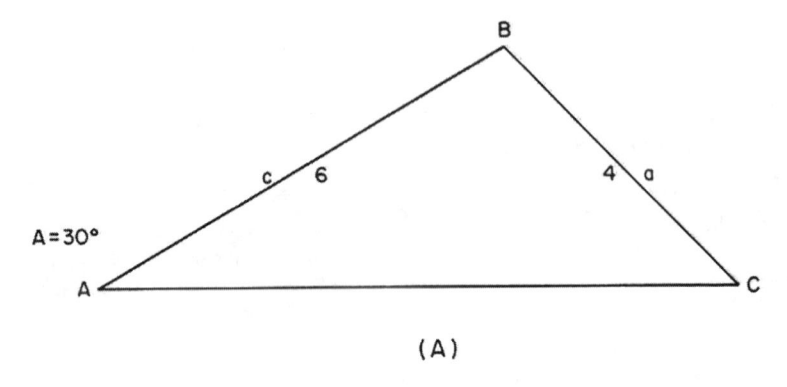

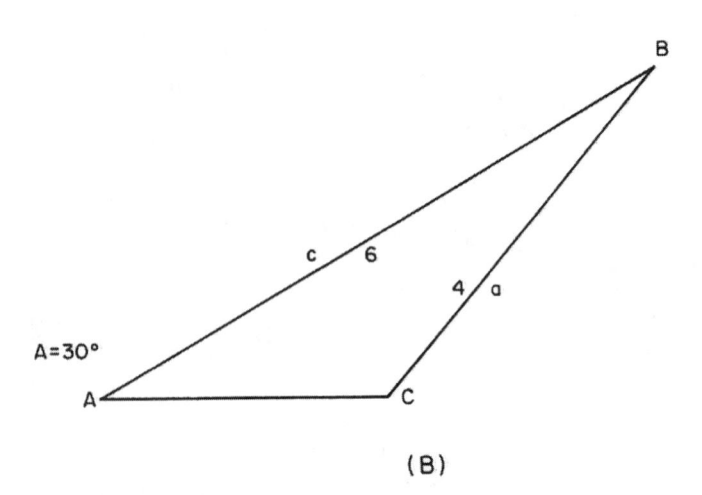

Figure 5-12.—Two different triangles derived from identical given data.

A = 22°, a = 5.4, c = 14, and C = 76° 13'

In A'B'C'; A' = A, a' = a, c' = c, and

Two Sides and an Opposite Angle

$C' = 103° 47'$.

Solving ABC first, angle B is found by

$$B = 180° - (A + C)$$

$$B = 180° - (22° + 76° 13')$$

$$B = 180° - 98° 13'$$

$$B = 81° 47'$$

Then by the law of sines

$$b = \frac{a \sin B}{\sin A}$$

$$b = \frac{5.4 \times 0.98973}{0.37461}$$

$$b = 14.27$$

This completes the solution of triangle ABC.
In the second triangle, angle B' is found first by using values previously calculated

$$B' = 180° - (A' + C')$$

$$B' = 180° - (22° + 103° 47')$$

$$B' = 180° - 125° 47'$$

$$B' = 54° 13'$$

Then from the law of sines

$$b' = \frac{a' \sin B'}{\sin A'}$$

$$b' = \frac{5.4 \times \sin 54° \ 13'}{\sin 22°}$$

$$b' = \frac{5.4 \times 0.81123}{0.37461}$$

$$b' = 11.70$$

This completes the solution for both possible triangles. Figure 5-13 shows a scale drawing of the triangle of this example. The angle A was constructed with one side of length 14 units terminating at B, and the other line to form an angle of 22°. A compass was set to 5.4 units and an arc was struck using B as the center. As shown in the figure, the arc intersected the line from A at two points. Therefore, there are two triangles which satisfy the data.

EXAMPLE: Solve the triangle (or triangles) when $A = 35°$, $a = 10$, $b = 8$.

SOLUTION: From the law of sines

$$\sin B = \frac{b \sin A}{a}$$

$$\sin B = \frac{8 \sin 35°}{10}$$

$$\sin B = 0.45886$$

Two Sides and an Opposite Angle

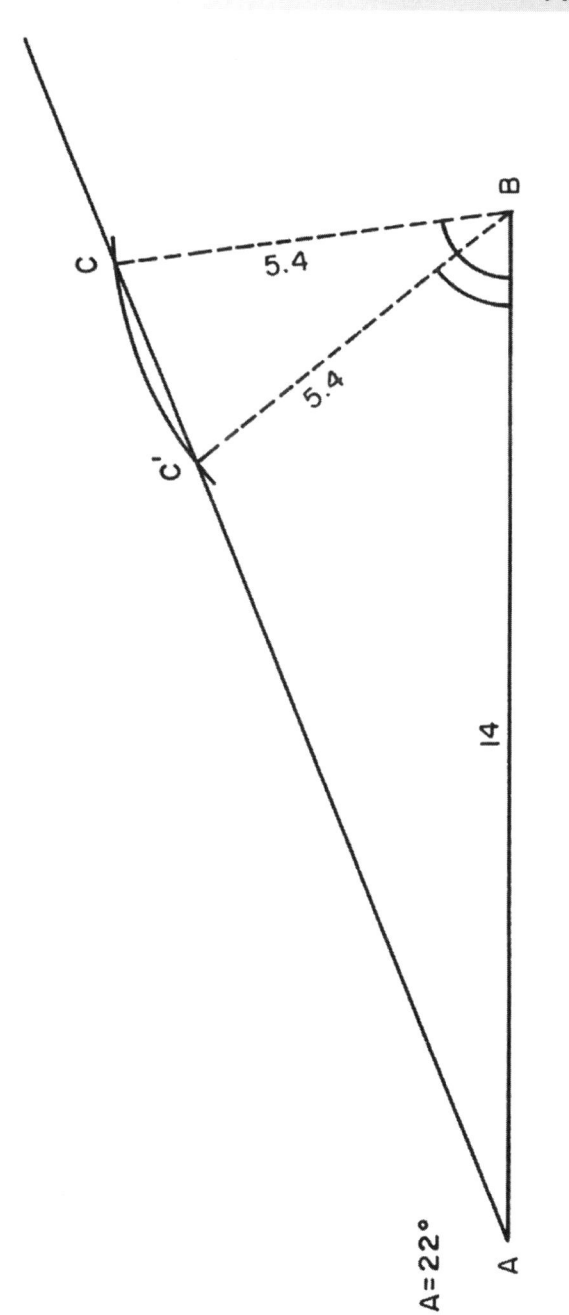

Figure 5-13.—Scale drawing of example triangles.

Then

$$B = 27° 19'$$

$$C = 180° - (A + B)$$

$$C = 180° - (35° + 27° 19')$$

$$C = 117° 41'$$

Apply the law of sines again to find c

$$c = \frac{a \sin C}{\sin A}$$

$$c = \frac{10 \sin 117° 41'}{\sin 35°}$$

$$c = \frac{10 \times 0.88553}{0.57358}$$

$$c = 15.44$$

This solves one triangle and since the side opposite the given angle is larger than the other given side, there should be only one solution. However, if this point is overlooked and the solution is continued in an attempt to find a second solution as in the previous example

$$\sin C' = \sin (180° - C)$$

$$\sin C' = \sin (180° - 117° 41')$$

$$\sin C' = \sin 62° 19'$$

Two Sides and an Opposite Angle

$$C' = 62° 19'$$

Now if this angle is contained in a second triangle described by the given data, it is known that

$$A + B + C' = 180°$$

but

$$35° + 27° 19' + 62° 19' = 124° 38'$$

so

$$A + B + C' \neq 180°$$

and ABC' is not a triangle described by the given data.

Practice Problems

In the following problems use the given data to solve the triangle or triangles involved.

1. $C = 100°$, $c = 46$, $b = 30$
2. $A = 40°$, $a = 25$, $b = 30$
3. $B = 42°$, $b = 2$, $c = 4$
4. $B = 30°$, $b = 10$, $a = 10$

Answers

1. $B = 39° 58'$, $A = 40° 02'$, $a = 30.04$
2. $B = 50° 29'$, $C = 89° 31'$, $c = 38.89$ and $B = 129° 31'$, $C = 10° 29'$, $c = 7.08$
3. No solution
4. $A = 30°$, $C = 120°$, $c = 17.32$

Logarithmic Solutions

Computations involving triangle solutions are often concerned with the multiplication or division of trigonometric functions, which contain values given in four or five decimal places, and other values which may also contain numerous digits. There are many opportunities for arithmetic errors in these computations and, in many cases, the errors are the result of the multiplication and division by large decimals. In logarithmic solution the computations are reduced to addition problems, and the number of errors is frequently reduced.

By the combined use of tables of trigonometric functions and tables of logarithms one could solve the triangles by first finding the value for the function and converting this to a logarithm or by converting the logarithm of a function value to a decimal and then converting this to an angle. However, the logarithm equivalents of the principal natural functions have long since been worked out, and are available in tables, so that the necessary multiplication or division may be performed by the use of logarithms.

A table of "common logarithms of trigonometric functions" usually lists the logs for the sine, cosine, tangent, and cotangent of angles from $0°$ through $180°$. Appendix I shows a sample page from such a table; reference to the appendix shows that both the characteristic and the mantissa are listed. In addition, for each value listed, a characteristic of -10 at the end of the log is understood.

Logarithmic Solutions

Take the log listed for the sine of 38° 00' 00", for example. This is listed as 9.78934. What this actually means is 9.78934-10, which in turn means that the log of this function is actually -1+.78934. On the other hand, the log listed for the tangent of 51° 10' 00" is 10.09422. What this means is 10.09422-10; in other words, the log of this function is 0.09422. The logs are printed in this manner simply to avoid the necessity for printing minus characteristics. Note that, even when a characteristic is minus, the mantissa is considered as plus.

A complete table of the logarithms of trigonometric functions is not included in this course. For purposes of the examples and practice problems in this course, a short table of values for the sine, cosine, and tangent is given in table 5-1. The values in this table are given for each 5° from 0° to 90°. The complete logarithm (both characteristic and mantissa) is given in this table. The problems in this section will be worked with an accuracy of side lengths to three digits and angles to the nearest 5°.

The solution of oblique triangles was considered in four cases. In logarithmic solutions the solutions are also considered in four cases. In cases 1 and 4 the law of sines was used for solutions. Since the law of sines fits well with logarithmic solutions, cases 1 and 4 also use the law of sines for logarithmic solutions. Cases 2 and 3 were solved using the law of cosines; however, the law of cosines involves addition and subtraction and does not lend itself to logarithmic solutions. For cases 2 and 3, we will use methods other than the law of cosines for logarithmic solutions.

Cases 1 and 4

As previously stated, the solution of these two cases by the law of sines adapts readily to logarithmic solution since the law of sines involves multiplication and division. Example solutions of the cases are given in the following paragraphs.

EXAMPLES: Use logarithms to solve the triangle ABC when $A = 110°$, $B = 25°$, and $c = 125$.

SOLUTION: Find the unknown angle as the first step

$$C = 180° - A - B$$

$$C = 180° - 110° - 25°$$

$$C = 45°$$

Using the law of sines with ratios involving a and c find the value of a as follows:

$$\frac{a}{\sin A} = \frac{c}{\sin C}$$

$$a = \frac{c}{\sin C} \times \sin A$$

Taking the logarithm of both sides of the equation gives

$$\log a = (\log c - \log \sin C) + \log \sin A$$

The logarithms of trigonometric functions given in table 5-1 include only angles from 0° to 90° so the equation above becomes

log a = (log c - log sin C) + log sin (180° - A)

log a = (log 125 - log sin 45°) + log sin 70°

Refer to table 5-1 and convert the values to logarithms. One method of simplifying the computation is to convert each logarithm to one with an end characteristic of -10 and use the following procedure

$$\begin{array}{ll} \log 125 & = 12.0969 - 10 \\ \log \sin 45° & = \underline{9.8495 - 10} \end{array} \quad \text{subtract}$$

log (c/sin C) = 2.2474

log (125/sin 45°) = 12.2474 - 10   add

log sin 70°   = $\underline{9.9730 - 10}$

log a   = 22.2204 - 20

log a = 2.2204

Taking the antilog,

a = 166

To complete the solution, find side b using the same procedure

$$\frac{b}{\sin B} = \frac{a}{\sin A}$$

$$b = \frac{a}{\sin A} \times \sin B$$

45

Math Made Nice-n-Easy

Table 5-1.—Logarithms of trigonometric functions.

Degrees	Log sin	Log cos	Log tan
0	—	.0000	—
5	8.9403 - 10	9.9983 - 10	8.9420 - 10
10	9.2397 - 10	9.9934 - 10	9.2463 - 10
15	9.4130 - 10	9.9849 - 10	9.4281 - 10
20	9.5341 - 10	9.9730 - 10	9.5611 - 10
25	9.6260 - 10	9.9573 - 10	9.6687 - 10
30	9.6990 - 10	9.9375 - 10	9.7614 - 10
35	9.7586 - 10	9.9134 - 10	9.8452 - 10
40	9.8081 - 10	9.8843 - 10	9.9238 - 10
45	9.8495 - 10	9.8495 - 10	0.0000
50	9.8843 - 10	9.8081 - 10	0.0762
55	9.9134 - 10	9.7586 - 10	0.1548
60	9.9375 - 10	9.6990 - 10	0.2386
65	9.9573 - 10	9.6260 - 10	0.3313
70	9.9730 - 10	9.5341 - 10	0.4389
75	9.9849 - 10	9.4130 - 10	0.5720
80	9.9934 - 10	9.2397 - 10	0.7537
85	9.9983 - 10	8.9403 - 10	1.0580
90	.0000	—	—

$\log b = (\log a - \log \sin A) + \log \sin B$

$\log b = (\log 166 - \log \sin 70°) + \log \sin 25°$

$\log 166$	$= 12.2204 - 10$	
$\log \sin 70°$	$= \underline{9.9730 - 10}$	subtract
$\log (a/\sin A)$	$= 2.2474$	

$\log (a/\sin A)$	$= 12.2474 - 10$	
$\log \sin 25°$	$= \underline{9.6260 - 10}$	add
$\log b$	$= 21.8734 - 20$	

$\log b = 1.8734$

$b = 74.7$

This completes the logarithmic solution for a triangle in case 1.

EXAMPLE: Solve the triangle (or triangles) when $A = 40°$, $a = 3$, $b = 4$.

This is an ambiguous case and with the side opposite the given angle smaller than the other given side there are two possibilities: either there are two solutions or side a is too short to reach the baseline and there are no solutions. Recall from earlier examples that there is no solution when the sine of the angle opposite the second given side (angle B in this case) is greater than one. In logarithmic solutions the condition for no solution is when the log sin of the angle is greater than zero; this corresponds to a sine greater than one. Reference to table 5-1 shows that all of the values listed for log sin are less than zero (negative characteristic).

SOLUTION: Solve first for the unknown angle opposite a given side using the law of sines

$$\sin B = \frac{b \sin A}{a}$$

this in logarithmic form becomes

$$\log \sin B = \log b + \log \sin A - \log a$$

Evaluation of the logarithms gives

log b	= 10.6021 - 10	
log sin A	= 9.8081 - 10	add
log (b x sin A)	= 20.4102 - 20	
log (b x sin A)	= 20.4102 - 20	
log a	= 10.4771 - 10	subtract
log sin B	= 9.9331 - 10	

We note that log sin B is less than 0, so there are two angles to consider, say B and B', where sin (180° - B) = sin B'.

Now

$$\log \sin B = 9.9331 - 10$$

$$B = 60°$$

$$B' = 120°$$

Then the corresponding angles, C and C', are

$$C = 180° - (A + B)$$

$$C = 180° - (40° + 60°)$$

$$C = 80°$$

and

$$C' = 180° - (A + B')$$

$$C' = 180° - (40° + 120°)$$

$$C' = 20°$$

To complete the solution find sides c and c'. Consider first side c and use the following form of the law of sines

$$c = \frac{a \sin C}{\sin A}$$

or, in logarithmic form,

$$\log c = \log a + \log \sin C - \log \sin A$$

Evaluate the logarithms

$\log a$	=	10.4771 - 10	
$\log \sin C$	=	9.9934 - 10	add
$\log (a \times \sin C)$	=	20.4705 - 20	
$\log \sin A$	=	9.8081 - 10	subtract
$\log c$	=	10.6624 - 10	

Taking the antilog,

$$c = 4.6$$

Finally, to find c' use

$$c' = \frac{a \sin C'}{\sin A}$$

$\log c' = \log a + \log \sin C' - \log \sin A$

$\log a$	=	10.4771 - 10	
$\log \sin C'$	=	9.5341 - 10	add
$\log (a \times \sin C')$	=	20.0112 - 20	
$\log (a \times \sin C')$	=	10.0112 - 10	subtract
$\log \sin A$	=	9.8081 - 10	
$\log c'$	=	0.2031	

Therefore,

$$c' = 1.6$$

Practice Problems

Use logarithms to solve the triangle (or triangles) described by the following data.

1. $A = 70°$, $B = 100°$, $c = 50$
2. $A = 60°$, $a = 11$, $b = 18$
3. $A = 40°$, $a = 25$, $b = 30$

Answers

1. $C = 10°$, $a = 271$, $b = 284$
2. No solution
3. $B = 50°$, $C = 90°$, $c = 39.8$
4. $B' = 130°$, $C' = 10°$, $c = 6.75$

Law of Tangents

The law of cosines does not lend itself to logarithmic solutions. The two cases in which we used the law of cosines are solved by logarithms using two different methods. The first method to solve triangles in case 3, where two sides and the included angle are given.

The law of tangents is expressed in words as follows:

In any triangle the difference between two sides is to their sum as the tangent of half the difference of the opposite angles is to the tangent of half their sum.

For any pair of sides—such as side a and side b—the law may be expressed as follows:

$$\frac{a - b}{a + b} = \frac{\tan 1/2 \, (A - B)}{\tan 1/2 \, (A + B)}$$

The law may be expressed in a form that includes other combinations of sides and angles by systematically changing the letters in the formula.

In solving case 3 by the law of tangents, select the formula which includes the given sides, say a and b; then angle C is also given. The sum of the unknown angles, A + B, is found as

$$A + B = 180° - C$$

and the law of tangents is used to find A - B. After the sum and difference of A and B are determined, the angles themselves can be found. With the angles known, the law of sines is used to find the unknown side.

EXAMPLE: Solve the triangle ABC when A = 25°, b = 10, c = 7.

SOLUTION: With b and c given and b > c use the law of tangents in the form

$$\frac{b - c}{b + c} = \frac{\tan 1/2 \, (B - C)}{\tan 1/2 \, (B + C)}$$

First, determine the sum of the unknown angles

$$B + C = 180° - A$$

$$B + C = 180° - 25°$$

$$B + C = 155°$$

Then $\quad 1/2(B + C) = 1/2(155°)$

$$1/2(B + C) = 77.5°$$

Law of Tangents

and

$$b - c = 10 - 7 = 3$$

$$b + c = 10 + 7 = 17$$

At this point, the only unknown in the formula for the law of tangents is $\tan 1/2(B - C)$. The next step is to transpose the formula and solve for this unknown.

$$\frac{b - c}{b + c} = \frac{\tan 1/2(B - C)}{\tan 1/2(B + C)}$$

$(b - c)(\tan 1/2(B + C)) = (\tan 1/2(B - C))(b + c)$

$$\frac{(b - c)(\tan 1/2(B + C))}{b + c} = \tan 1/2(B - C)$$

$$\tan 1/2(B - C) = \frac{3 \tan 77.5°}{17}$$

Rounding the angle to 80° for use with the given table, the following logarithmic equation can be written:

$\log \tan 1/2(B - C) = \log 3 + \log \tan 80° - \log 17$

then

$$\log 3 = 10.4771 - 10$$
$$\log \tan 80° = 10.7537 - 10 \quad (+)$$

$$\log (3 \times \tan 80°) = 21.2308 - 20$$
$$\log 17 = 11.2304 - 10 \quad (-)$$
$$\log \tan 1/2(B - C) = 10.0004 - 10$$
$$\log \tan 1/2(B - C) = 0.0004$$
$$1/2(B - C) = 45°$$
$$B - C = 90°$$

There are now two equations for B and C, (B + C) and (B - C); these are solved simultaneously to find B and C.

First the two are added to find B

$$B + C = 155°$$
$$\underline{B - C = 90°}$$
$$2B = 240°$$

$$B = 120°$$

Next, subtract the two to find C

$$B + C = 155°$$
$$\underline{-B + C = -90°}$$
$$2C = 65°$$

54

Law of Tangents

$$C = 32.5°$$

To fit the given table round C to 35°, then

$$A + B + C = 25° + 120° + 35° = 180°$$

In the final part of the solution, use the law of sines to find side a.

$$\frac{a}{\sin A} = \frac{b}{\sin B}$$

$$a = \frac{b \sin A}{\sin B}$$

$\log a = \log b + \log \sin A - \log \sin B$

$\log a = \log 10 + \log \sin 25° - \log \sin 120°$

$$\log 10 = 11.0000 - 10$$
$$\log \sin 25° = \underline{9.6260 - 10} \quad (+)$$

$$\log (b \sin A) = 20.6260 - 20$$
$$\log \sin B = \underline{9.9375 - 10} \quad (-)$$

$$\log a = 10.6885 - 10$$

$$\log a = 0.6885$$

$$a = 4.88$$

There are six forms of the law of tangents, two of which have been shown. The remaining four are formed, as previously stated, by replacing the corresponding letters in the formula.

Practice Problems

Use logarithms and the law of tangents to solve the triangles described by the given data.
1. $a = 4$, $b = 3$, $C = 60°$
2. $a = 0.0316$, $b = 0.0132$, $C = 50°$

Answers

1. $A = 70°$, $B = 50°$, $c = 3.69$
2. $A = 105°$, $B = 25°$, $c = 0.0239$

Case 2

The logarithmic solution of oblique triangles when three sides are given involves formulas derived from the law of cosines. These formulas are called half-angle formulas and, in these solutions, are expressed in terms of the semiperimeter of the triangle and the radius of a circle inscribed in the triangle.

The half-angle formulas expressed in s (semiperimeter) and r (radius of inscribed circles) are not presented in this course. The logarithmic solutions of oblique triangles in this course are limited to cases 1, 3, and 4.

Area Formula

In this section two formulas for finding the area of oblique triangles are given. These formulas are used to find the area of triangles in cases 1 and 3 from the given parts.

Case 2

Recall from plane geometry that the area of a triangle is found by the formula

$$A = \frac{1}{2} bh$$

where b is any side of the triangle and h is the altitude drawn to that side. To avoid confusion between A for area and A as an angle in the triangle the word "area" will be used in this chapter. Then the formula is stated as

$$\text{area} = \frac{1}{2} bh$$

Reference to figure 5-3 (B) shows that the length of the altitude h can be found by

$$h = c \sin A$$

Substituting this value of h in the area formula results in

$$\text{area} = \frac{1}{2} bc \sin A$$

$$\text{area} = \frac{bc \sin A}{2}$$

This area formula is stated in words as follows: the area of a triangle is equal to one-half the product of two sides and the sine of the angle between them. This formula is used for solutions of triangles when two sides and the included angle are given.

A second formula for area can be derived from the law of sines and the previous formula. From the law of sines

$$\frac{b}{\sin B} = \frac{c}{\sin C}$$

$$b = \frac{c \sin B}{\sin C}$$

Substituting this value of b in the area formula

$$\text{area} = \frac{bc \sin A}{2}$$

results in

$$\text{area} = \left(\frac{c \sin B}{\sin C}\right)\left(\frac{c \sin A}{2}\right)$$

$$\text{area} = \frac{c^2 \sin A \sin B}{2 \sin C}$$

This formula can be used to solve triangles when two angles and one side are given since, when two angles are given, the third angle can be found directly. It can be seen that the area formulas can be easily adapted to logarithmic solutions, as well as to normal solutions.

Practice Problems

Derive the area formulas most applicable when the following are given.

1. a, b, C
2. a, c, B
3. A, C, b
4. B, C, a

Area Formula

Answers

1. area $= \dfrac{ab \sin C}{2}$

2. area $= \dfrac{ac \sin B}{2}$

3. area $= \dfrac{b^2 \sin A \sin C}{2 \sin B}$

4. area $= \dfrac{a^2 \sin B \sin C}{2 \sin A}$

EXAMPLE: Find the area of triangle ABC when A = 40°, b = 13, and c = 9.

SOLUTION: Use the first area formula given in this section:

$$\text{area} = \frac{bc \sin A}{2}$$

$$\text{area} = \frac{13 \times 9 \times \sin 40°}{2}$$

$$\text{area} = \frac{13 \times 9 \times 0.64279}{2}$$

$$\text{area} = 37.6 \text{ (square units)}$$

Math Made Nice-n-Easy

This formula (as well as the other area formula) adapts easily to logarithmic solutions. In a logarithmic solution the formula

$$\text{area} = \frac{bc \sin A}{2}$$

is written

$$\log \text{area} = \log b + \log c + \log \sin A - \log 2$$

EXAMPLE: Find the area of triangle ABC when $A = 25°$, $B = 105°$, $c = 12$.

SOLUTION: First determine angle C.

$$C = 180° - (A + B)$$
$$C = 180° - 130°$$
$$C = 50°$$

then apply the area formula

$$\text{area} = \frac{c^2 \sin A \sin B}{2 \sin C}$$

$$\text{area} = \frac{(12)^2 \sin 25° \sin 105°}{2 \sin 50°} \quad (1)$$

$$\text{area} = \frac{144 \times 0.42262 \times 0.96593}{2 \times 0.76604}$$

With the specific values involved in this problem, a logarithmic solution should simplify the arithmetical process. To write the logarithmic equation go back to equation (1). Recall that the logarithm of 12^2 is $2 \log 12$ and that $\sin(180° - 105°) = \sin 75°$ and write the equation as

Area Formula

$$\log \text{area} = 2 \log 12 + \log \sin 25°$$
$$+ \log \sin 75° - (\log 2 + \log 50°)$$

$$2 \log 12 = 2.1584$$
$$\log \sin 25° = 9.6260 - 10$$
$$\log \sin 75° = \underline{9.9849 - 10} \quad \text{add}$$
$$\log (c^2 \sin A \sin B) = 21.7693 - 20$$
$$\log (c^2 \sin A \sin B) = 11.7693 - 10$$

and

$$\log 2 = 0.3010$$
$$\log \sin 50° = \underline{9.8843 - 10} \quad \text{add}$$
$$\log (2 \sin C) = 10.1853 - 10$$

subtracting these sums

$$\log (c^2 \sin A \sin B) = 11.7693 - 10$$
$$\log (2 \sin C) = \underline{10.1853 - 10} \quad \text{subtract}$$
$$\log \text{area} = 1.5840$$
$$\text{area} = 38.37$$

Math Made Nice-n-Easy

Practice Problems

Find the area of the triangles described by the given data.

1. Find the area of the triangle given in the second example problem (A = 25°, B = 105°, c = 12), without using logarithms.

2. A = 25°, B = 45°, c = 24

3. A = 42°, b = 4.4, c = 3

4. A = 120°, b = 8, c = 12

Answers

1. 38.369
2. 91.6
3. 4.4
4. 41.5

Chapter 6
Vectors

Vector quantities, which we will now concern ourselves with, are different from scalar quantities. Navigation involves the use of vector quantities, surveyors use vectors in their work, as do structural engineers, and electrical and electronic technicians. Many of the applications of electricity and electronics involve the use of vector quantities.

Definitions and Terms

In this section we will make a distinction between a scalar quantity and a vector. We will also define the coordinate systems used in working with vectors and show some of the symbols used.

Scalars

Heretofore, we have been concerned with scalar quantities, which are measurements or quantities having only magnitude, in the appropriate units. Examples of scalar quantities are: 10 pounds, 4 miles, 17 feet, and 28.2 pounds per square inch.

Vectors

A vector, in contrast to a scalar, has direction as well as magnitude. Examples of vector quantities are: 6 miles due north, 9 blocks toward the west, and 250 knots at 30°. Notice that the vector quantities have both a magnitude and a direction.

Symbols

The letters A, B, C, and D have been previously used to represent scalar quantities in algebra. In vector algebra, a notation is used to denote scalar symbols in relation to vector symbols. A dash over the letters, for example $\overline{A}$ and $\overline{B}$, denotes vectors.

A vector can conveniently be represented by a straight line. The length of this straight line represents the magnitude, and its position in space represents the direction of the vector quantity. In figure 6-1 the vectors $\overline{A}$, $\overline{B}$, and $\overline{C}$ are equal because they have the same magnitude and direction and vector $\overline{D}$ is not equal to either vector $\overline{A}$, $\overline{B}$, or $\overline{C}$ because although it has the same magnitude it does not have the same direction.

In navigational problems, a coordinate system is used in which the compass points serve as indicators of direction, and magnitude is given by lengths of the lines. For example, using the origin of the coordinate system as the point of departure from the harbor, figure 6-2 represents two ships heading out to sea. Vector $\overline{A}$ represents a ship bearing 45° from due north at a speed of 20 knots, and vector $\overline{B}$

Symbols

represents a ship bearing 60° from due north at a speed of 25 knots. Notice that directions in this coordinate system are measured clockwise from due north.

When we use the trigonometric system in designating angles or giving a direction to a magnitude we use the Cartesian coordinate system which includes the abscissa (measurement on the X axis) and the ordinate (measurement on the Y axis). Directions in this coordinate system are measured counterclockwise from the X axis. For example, using the origin of the coordinate system as the point of departure from the harbor, figure 6-3 represents a ship heading out to sea bearing 30° at a speed of 25 knots. This is represented by vector $\overline{B}$ and is the same representation as vector $\overline{B}$ in figure 6-2 but is shown on a different coordinate system.

Angle measurements will be referenced from the vertical in the compass coordinate system and will be referenced from the horizontal in the Cartesian coordinate system.

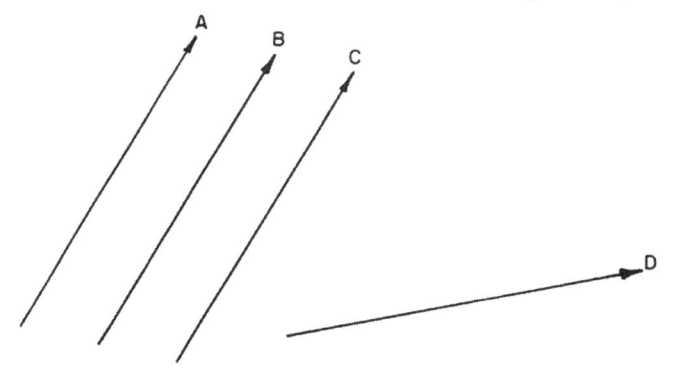

Figure 6-1.—Comparison of vectors.

Math Made Nice-n-Easy

Figure 6-2.—Compass coordinate system.

Figure 6-3.—Cartesian coordinate system.

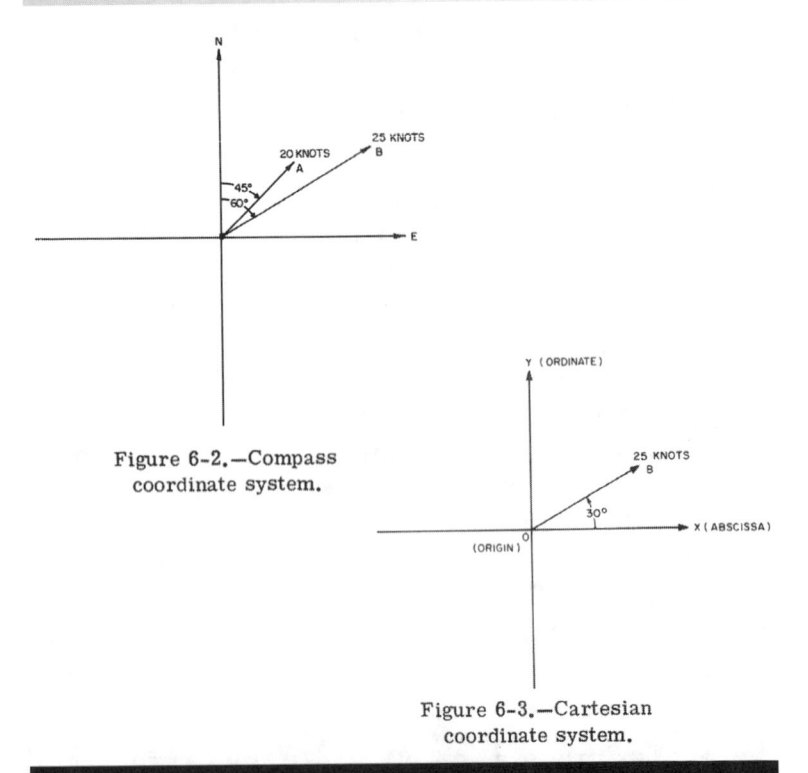

Combining Vectors

If a vector $\overline{A}$ represents the displacement of a particle or the force acting on the particle, it is convenient to let $-\overline{A}$ represent the displacement of a particle in the opposite direction or to represent a force in the direction opposite to $\overline{A}$. Thus, vectors $\overline{A}$ and $-\overline{A}$ are equal in magnitude but are opposite in direction.

Addition

The resultant of two vectors acting in the same direction or acting in opposite directions is the algebraic sum of their magnitudes. An

Addition

example of this is walking due east four steps and then walking due west one step. The resultant is three steps due east. If one travels from his home to his place of employment he may have several choices for his route. We will assume two of these routes as indicated in figure 6-4. He may move east to point A then north to point B or he may move north to point C then east to point B. In either case he arrives at his place of employment. If we assume his travel in both directions as forces acting on him, we can call the direct distance from home to place of employment, in figure 6-4, the RESULTANT of these two forces and refer to it as vector $\overline{R}$. Notice that either path taken results in vector $\overline{R}$.

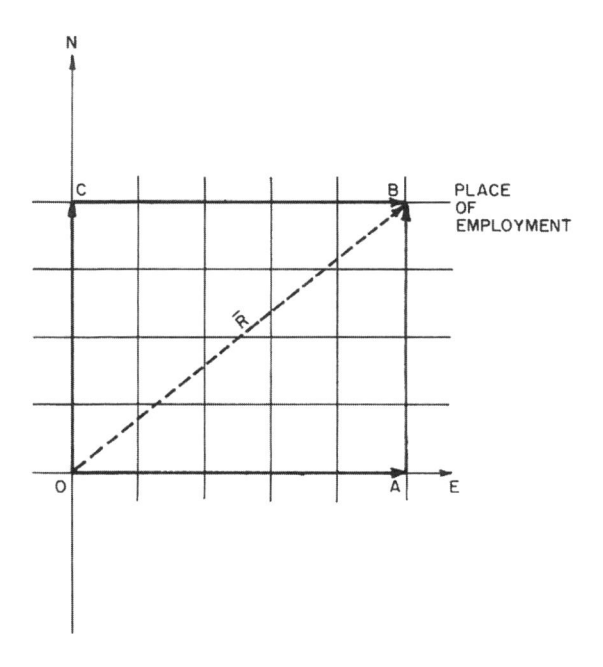

Figure 6-4.—Right angle vectors.

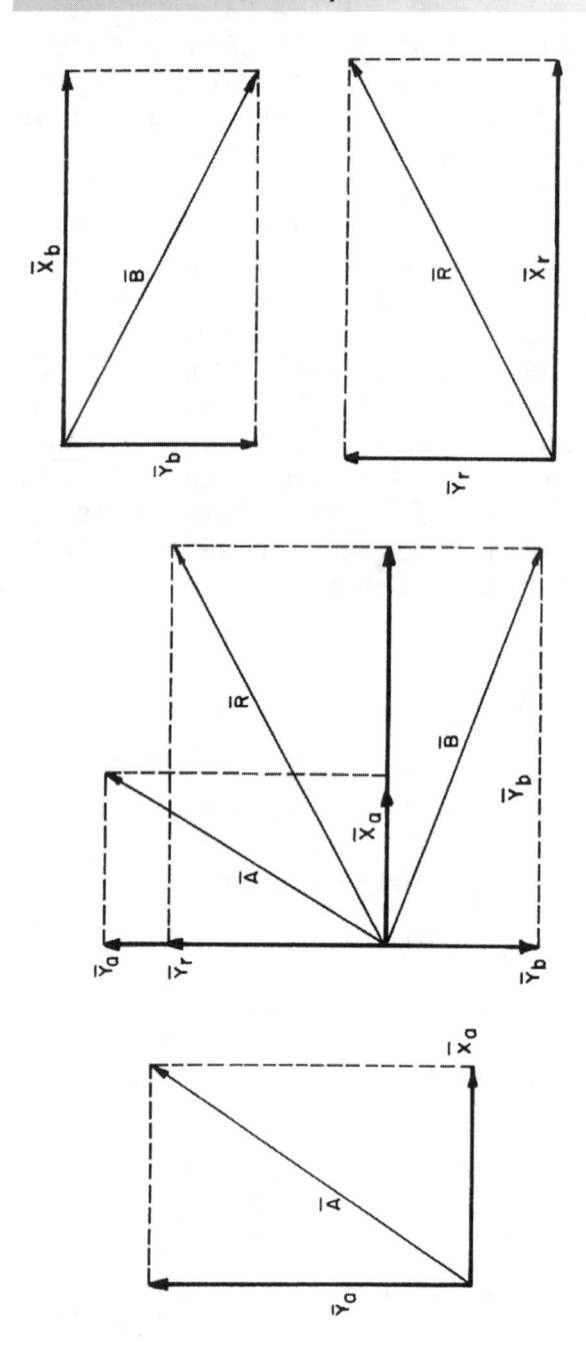

Figure 6-5.—Components of vectors.

Subtraction

We may now state that $\overline{OA} \leftarrow \overline{AB} = \overline{OC} \leftarrow \overline{CB} = \overline{R}$. The symbol ($\leftarrow$) is used to indicate that vector $\overline{OA}$ is added vectorially to vector $\overline{AB}$. From this it is apparent that vectors may be added in either order with the same results.

If several vectors $\overline{A}$, $\overline{B}$, and $\overline{R}$ are to be resolved into components, $\overline{X}_r$, $\overline{X}_a$, $\overline{X}_b$, and $\overline{Y}_r$, $\overline{Y}_a$, $\overline{Y}_b$ are used to denote these components, as figure 6-5 portrays.

In the addition of vectors, the initial point of vector $\overline{B}$ must be placed directly on the terminal point of vector $\overline{A}$, and so on for any number of vectors. Then vector $\overline{R}$, which joins the initial point of vector $\overline{A}$ with the terminal point of the last vector $\overline{N}$, is the result of adding vectors $\overline{A}$ and $\overline{B}$ and $\overline{C}$ through $\overline{N}$ vectorially. Hence $\overline{A} \leftarrow \overline{B} \leftarrow \overline{C} \leftarrow \overline{D} \leftarrow \ldots \leftarrow \overline{N} = \overline{R}$. Here it may be shown that the commutative and associative principles apply, which means that it makes no difference which vector is used first and which order is followed when adding vectors.

Subtraction

Subtracting a vector is defined as adding a negative vector:

$$\overline{A} - \overline{B} = \overline{A} \leftarrow (-\overline{B})$$

It follows that if

$$\overline{A} \leftarrow \overline{B} = 0$$

Then

$$\overline{A} = -\overline{B}$$

A careful study of figure 6-4 will show that if

$$\overline{OA} + \overline{AB} = \overline{R}$$

Then

$$\overline{R} + (-\overline{AB}) = \overline{R} - \overline{AB} = \overline{OA}$$

Vector Solutions

In the previous sections we agreed that:

$$\overline{A} + \overline{B} + \overline{C} + \overline{D} + \ldots + \overline{N} = \overline{R}$$

Approaching this relation from a graphical standpoint, one can understand exactly what this means.

Graphical

As an example of the graphical method, six vectors may be used to represent the path taken by a man looking for a lost golf ball. He stands at position P_0, hits the ball and does not notice where the ball went. Vectors $\overline{A}$ through $\overline{F}$ in figure 6-6 represent the path he takes in an attempt to find the ball. P_f is the position where the ball is found and we will call it the termination point. Figure 6-7 shows another of the many different arrangements of the six vectors. The dotted lines from P_0 to P_f indicate the resultant vector, and this resultant has the same magnitude and direction regardless of the arrangement of the vectors we use. This method of graphically solving vector

Graphical

problems is called the polygon method, and is used in civil engineering problems involving structures such as bridges; it is also used in logic problems of everyday living.

If two vectors are to be resolved into a single resultant, this may be done graphically by the parallelogram method. Given any two vectors $\overline{A}$ and $\overline{B}$ lying in a plane (fig. 6-8) form a parallelogram by projecting $\overline{B}$ onto $\overline{A}$, initial point to terminal point, and $\overline{A}$ onto $\overline{B}$, initial point to terminal point, thus forming a parallelogram which has as a diagonal the resultant vector $\overline{R}$.

This process can be reversed in order to find the components of a vector as shown in figure 6-9. Vector $\overline{R}$ is given and the problem is to find the rectangular components of this

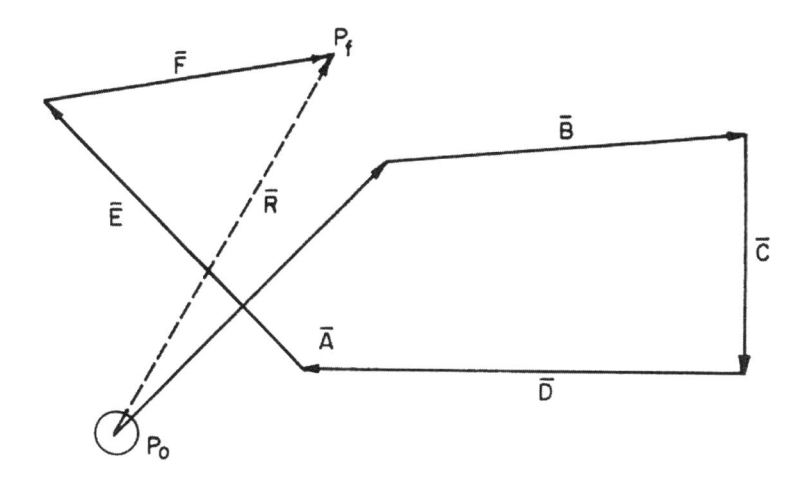

Figure 6-6.—Polygon example number 1.

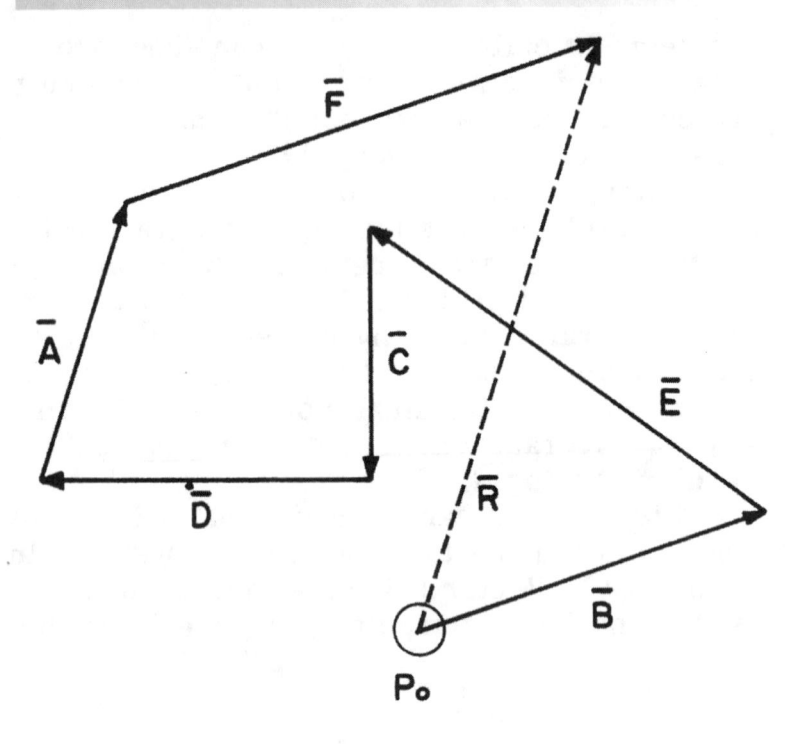

Figure 6-7.—Polygon example number 2.

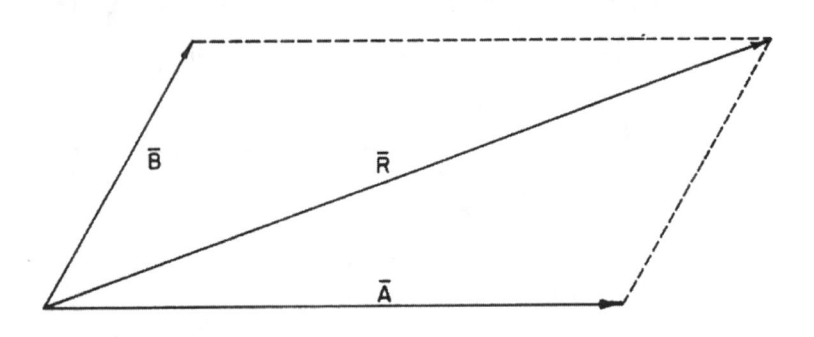

Figure 6-8.—Parallelogram example 1.

Analytical

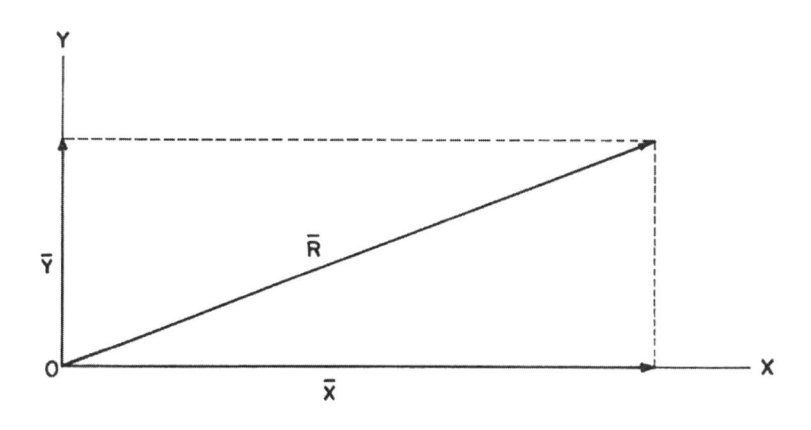

Figure 6-9.—Parallelogram example 2.

vector. In this case the parallelogram is a rectangle and the projections of $\bar{R}$ on the X axis and the Y axis show the components. Generally, the graphical method of resolving vectors will be used to check the validity of an analytical method of solution.

Analytical

The trigonometric functions are used to solve vector problems analytically.

EXAMPLE: Find the resultant of two vectors at right angles to each other. Vector $\bar{A}$ represents 90 pounds of force and vector $\bar{B}$ represents 60 pounds of force.

SOLUTION: Vector $\bar{A}$ is directed vertically and $\bar{B}$ lies on the reference line, as shown in figure 6-10. In this case the angle θ is unknown and the resultant is required. The Pythagorean theorem is sufficient to solve for

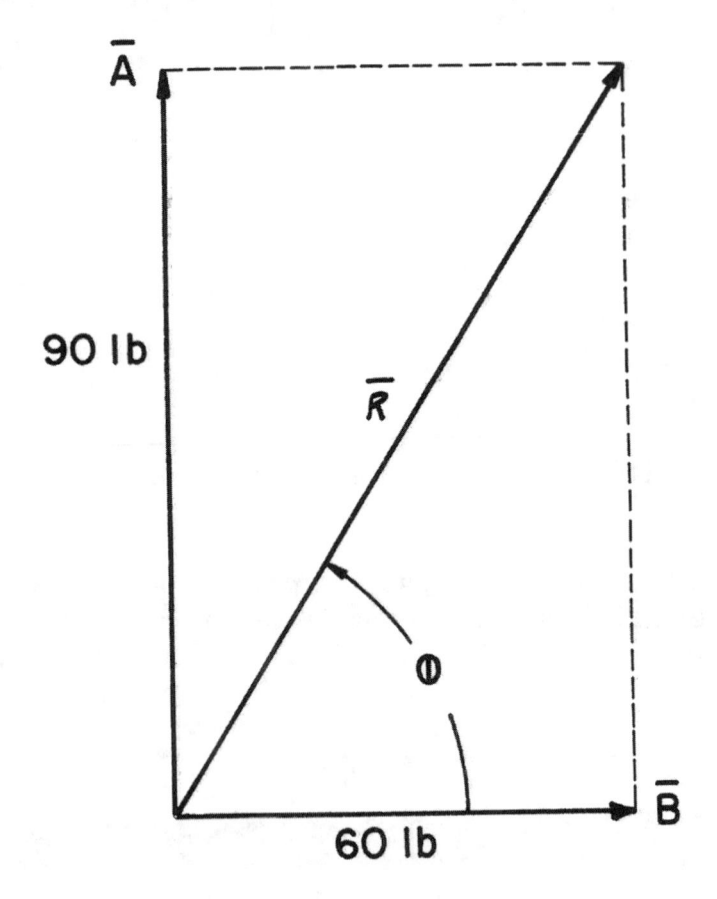

Figure 6-10.—Vector sum.

the magnitude of the resultant. This is the case only if we establish a right triangle from our vector and its components.

Since the magnitude of $\overline{R}$ is a scalar quantity we will designate it by r. As you recall, the Pythagorean theorem of right triangles states: $x^2 + y^2 = r^2$. We apply this to our figure and find that:

$$r^2 = (90)^2 + (60)^2$$

$$r = \sqrt{(90)^2 + (60)^2}$$

$$= \sqrt{11,700}$$

$$= 108.2 \text{ pounds}$$

If we desire the angle θ of the resultant vector $\overline{R}$ we may use the trigonometric function for the tangent of an angle; that is,

$$\tan \theta = \frac{y}{x} = \frac{90}{60} = 1.50000$$

Then,

$$\theta = 56° 19'$$

EXAMPLE: Resolve the vector $\overline{R}$ into its components. In figure 6-11, $\overline{R}$ represents 50 mph at 30°. We are to find the $\overline{X}_r$ and $\overline{Y}_r$ components.

SOLUTION: Recalling that the trigonometric functions for sin θ and cos θ are:

$$\sin \theta = \frac{y}{r}$$

and

$$\cos \theta = \frac{x}{r}$$

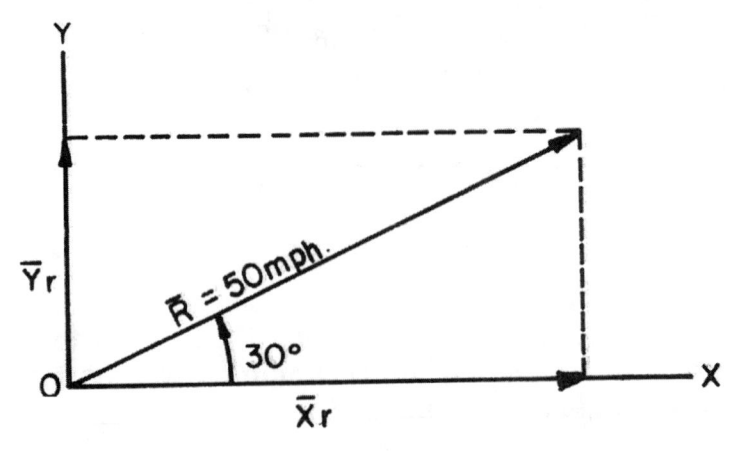

Figure 6-11.—Vector resolution.

we use these and find that

$$\sin 30° = \frac{y}{50}$$

then

$$50 \sin 30° = y$$
$$y = 25 \text{ mph}$$

and

$$\cos 30° = \frac{x}{50}$$

then

$$50 \cos 30° = x$$
$$x = 43.3 \text{ mph}$$

Analytical

Vector components acting in the same direction or in opposite directions may be added or subtracted algebraically. Vector components in the form $\overline{R} = \overline{X}_r + \overline{Y}_r$ fulfill this statement.

EXAMPLE: Add the following vectors given by their rectangular components.

SOLUTION:

If

$$\overline{A} = \overline{5} + \overline{2}$$

$$\overline{B} = \overline{6} - \overline{1}$$

Then $\quad \overline{A} + \overline{B} = \overline{11} + \overline{1}$

Observe that the first component of each pair is the X component, and the second is the Y component. If vector $\overline{A}$ is to be added to $\overline{B}$, the X component of the resultant is the sum of the X components of $\overline{A}$ and $\overline{B}$. The same reasoning applies to the Y components. Note especially that an X component is never added to a Y component or vice versa.

Practice Problems

Add the following rectangular form vectors.

1. $\overline{15} - \overline{5}$ and $\overline{3} + \overline{2}$
2. $\overline{3.96} + \overline{2.87}$ and $\overline{1.21} + \overline{3.11}$
3. $\overline{9.3} + \overline{4.8}$ and $\overline{0.2} - \overline{3.1}$
4. $\overline{182} + \overline{312}$ and $\overline{76} - \overline{81}$

Answers

1. $\overline{18} - \overline{3}$
2. $\overline{5.17} + \overline{5.98}$
3. $\overline{9.5} + \overline{1.7}$
4. $\overline{258} + \overline{231}$

EXAMPLE: Subtract the following vectors given by their rectangular components.

If

$$\overline{A} = \overline{4.2} + \overline{3.1}$$

$$\overline{B} = \overline{8.1} + \overline{6.2}$$

Then subtract $\overline{A}$ from $\overline{B}$

Thus

$$\overline{B} = \overline{8.1} + \overline{6.2}$$

$$\overline{A} = \overline{4.2} + \overline{3.1}$$

$$\overline{B} - \overline{A} = \overline{3.9} + \overline{3.1}$$

Practice Problems

Subtract the following rectangular form vectors.

1. Subtract $\overline{4.2} + \overline{3.1}$ from $\overline{8.1} + \overline{6.2}$
2. Subtract $\overline{57} + \overline{28}$ from $\overline{103} - \overline{35}$
3. Subtract $\overline{32.3} - \overline{8.3}$ from $\overline{15.3} + \overline{10.2}$
4. Subtract $\overline{-6.2} + \overline{2.9}$ from $\overline{-3.1} - \overline{2.6}$

Answers

1. $\overline{3.9} + \overline{3.1}$
2. $\overline{46} - \overline{63}$
3. $\overline{-17} + \overline{18.5}$
4. $\overline{3.1} - \overline{5.5}$

The notation up to this point has involved the regular rectangular coordinates. The form $\overline{R} = \overline{X}_r + \overline{Y}_r$ implies that a number of horizontal units and a number of vertical units combine to determine the end point of a vector. A second method commonly used describes a vector in terms of polar coordinates.

Polar Coordinates

If the length of the vector is known, all that is required to locate the vector is the angle through which it has been rotated. Measured from the reference line, the notation used is

$$\overline{R} = r \underline{/\theta}$$

where r is the magnitude and θ defines the direction. Thus, r is a scalar quantity. For instance, a vector 10 units long at 30° would be written $10\underline{/30°}$.

If $\overline{X}_r$ and $\overline{Y}_r$ are known, the scalar quantity r can be found by using the Pythagorean theorem:

$$r = \sqrt{(x_r)^2 + (y_r)^2}$$

The angle can then be found by using the tangent, thus:

Math Made Nice-n-Easy

$$\tan \theta = \frac{y_r}{x_r}$$

Now we have a method whereby we can change from the rectangular form to the polar coordinate form when working with vectors.

EXAMPLE: Change the vector $\overline{3} + \overline{4}$ into polar form. (NOTE: x_r is always placed first.)

SOLUTION:

$$r = \sqrt{3^2 + 4^2} = 5$$

$$\tan \theta = \frac{4}{3} = 1.33333$$

$$\theta = 53° \ 8'$$

If

$$\overline{R} = \overline{3} + \overline{4}$$

Then

$$\overline{R} = 5 \underline{/53° \ 8'}$$

Practice Problems

Change the rectangular form into polar form.
1. $\overline{1} + \overline{2}$
2. $\overline{8} + \overline{6}$
3. $\overline{\sqrt{3}} + \overline{1}$
4. $\overline{18.3} + \overline{2.8}$

Answers

1. 2.24/63° 26'
2. 10/36° 52'
3. 2/30°
4. 18.5/8° 42'

This method may be reversed and it is possible to change a vector from polar to rectangular coordinates.

EXAMPLE: Change the vector 30/65° into rectangular form.

SOLUTION:

If

$$\overline{R} = 30 \,/\, 65°$$

$$\overline{Y}_r = r \sin \theta$$

Then

$$\overline{Y}_r = 30 \sin 65°$$

$$= 27.18930$$

And

$$\overline{X}_r = 30 \cos 65°$$

$$\overline{X}_r = r \cos \theta$$

$$= 12.67860$$

Thus

$$\overline{R} = \overline{12.68} + \overline{27.19}$$

Practice Problems

Change the polar form to rectangular form.

1. $5/25°$
2. $83/72°$
3. $20/63°$
4. $8.2/31° \ 23'$

Answers

1. $\overline{4.53} + \overline{2.11}$
2. $\overline{25.6} + \overline{78.9}$
3. $\overline{9.08} + \overline{17.82}$
4. $\overline{7.00} + \overline{4.27}$

If we are to combine two vectors, proceed as follows:

EXAMPLE: Find the resultant of two vectors $\overline{A}$ and $\overline{B}$ if $\overline{A} = 12/102°$ and $\overline{B} = 5/12°$ in figure 6-12.

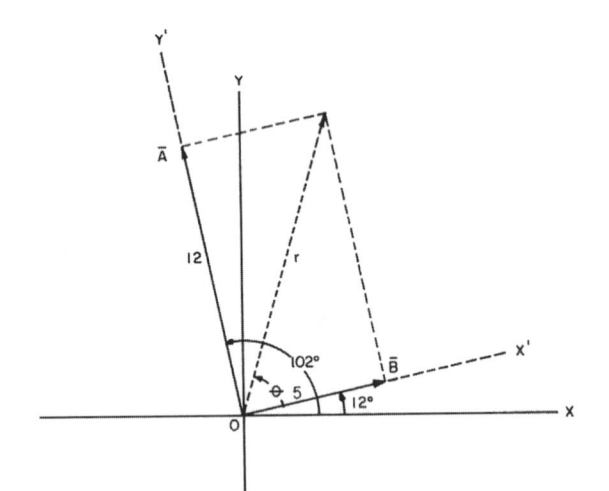

Figure 6-12.—A new reference.

Polar Coordinates

SOLUTION: In this problem we do not have either vector to coincide with the X axis or the Y axis. We may choose a new frame of reference, that is, X' and Y' to determine the magnitude of the resultant. Because $\overline{A}$ differs in direction by 90° from $\overline{B}$, we may still use the properties of right triangles.

Therefore $\quad r^2 = (12)^2 + (5)^2$

And $\quad r = \sqrt{(12)^2 + (5)^2}$

$\quad\quad\quad = \sqrt{169}$

$\quad\quad\quad = 13$

We now have the magnitude of our resultant and need only to find its direction. As we are concerned with only two vectors we may approach this problem in either of two ways. We may find our direction from our new reference then add the angle our new reference makes with the standard X axis and Y axis reference. In the new frame of reference we find the resultant to be:

$$\tan \theta = \frac{y'_r}{x'_r} = \frac{12}{5} = 2.40000$$

Therefore

$$\theta = 67° \; 23'$$

Now, the direction of the resultant is 67° 23' from the X' axis but the X' axis is 12° from the X axis so the resultant is 67° 23' + 12° or 79° 23' from the X axis.

Another approach to this problem is by resolving each of our vectors into their X axis and Y axis components and then adding these components algebraically. We have 5/12° which resolves into the following (fig. 6-13):

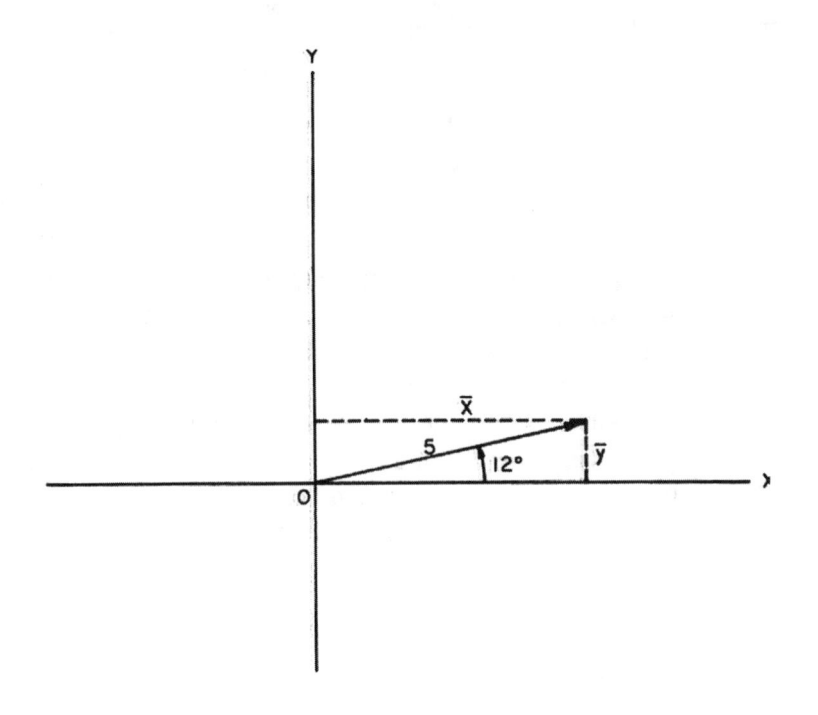

Figure 6-13.—Components for 5/12°.

$$\sin\theta = \frac{y}{r}$$

$$= \frac{y}{5}$$

Therefore

$$\bar{y} = 5 \sin 12°$$

$$= 5(0.20791)$$

$$= 1.03955$$

And

$$\cos\theta = \frac{x}{r}$$

$$= \frac{x}{5}$$

Therefore

$$\bar{x} = 5 \cos 12°$$

$$= 5(0.97815)$$

$$= 4.89075$$

We have now determined the X axis and Y axis components of one of the vectors. We will proceed to find the components of the other vector, 12/102°, as shown in figure 6-14. The Y axis component is as follows:

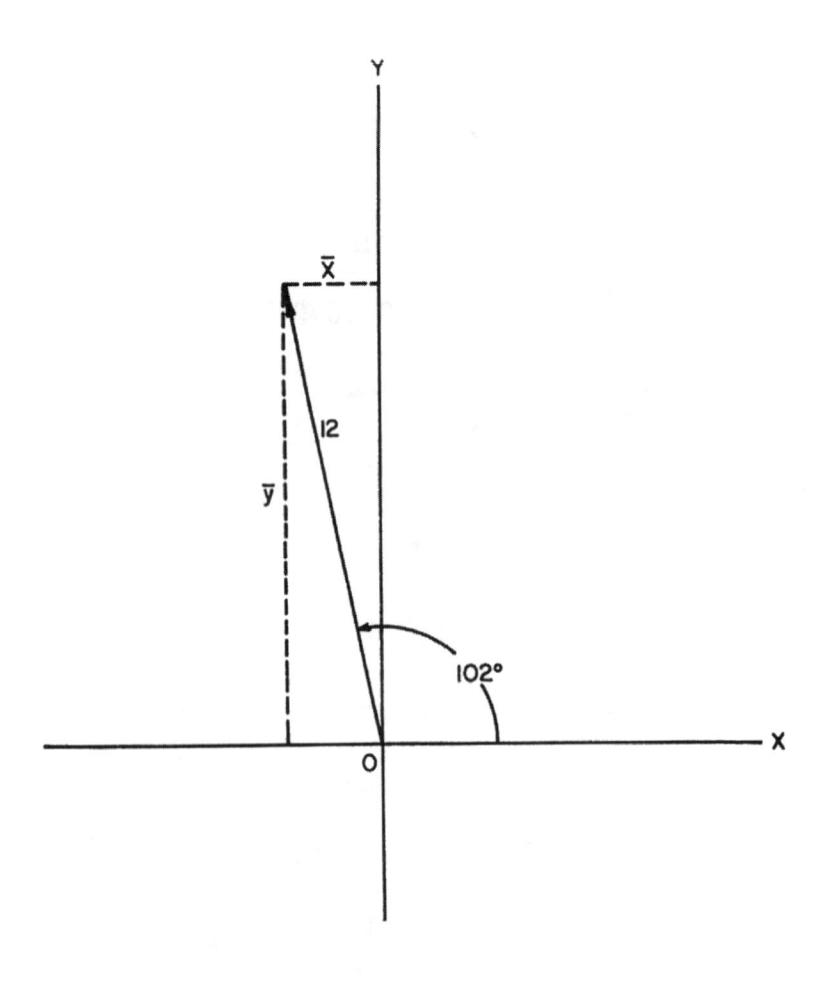

Figure 6-14.—Components for 12/102°.

If
$$\sin \theta = \frac{y}{r}$$
$$\sin \theta = \frac{y}{12}$$

Then
$$\overline{Y} = 12 \sin 102°$$

And, since
$$\sin (180° - \theta) = \sin \theta$$

Then
$$\sin 102° = \sin (180° - 78°)$$
$$= \sin 78°$$

Therefore
$$\overline{Y} = 12 \sin 78°$$
$$= 12(0.97815)$$
$$= 11.73780$$

We now find the X axis components as follows:
If
$$\cos \theta = \frac{x}{r}$$
$$\cos \theta = \frac{x}{12}$$

Then
$$\overline{X} = 12 \cos 102°$$

And, since
$$\cos(180° - \theta) = -\cos\theta$$

Then
$$\cos 102° = \cos(180° - 78°)$$
$$= -\cos 78°$$

Therefore
$$\overline{X} = 12(-0.20791)$$
$$= -2.49492$$

We must now add the X axis components and the Y axis components of the two vectors algebraically as follows:

$$\overline{Y} = 1.03955 + 11.73780$$
$$= 12.77735$$

And
$$\overline{X} = 4.89075 + (-2.49492)$$
$$= 2.39583$$

We now have $\overline{x}_r$ and $\overline{y}_r$ in rectangular form and may use the Pythagorean theorem to find the resultant in scalar measurement as shown in figure 6-15. This is as follows:

Polar Coordinates

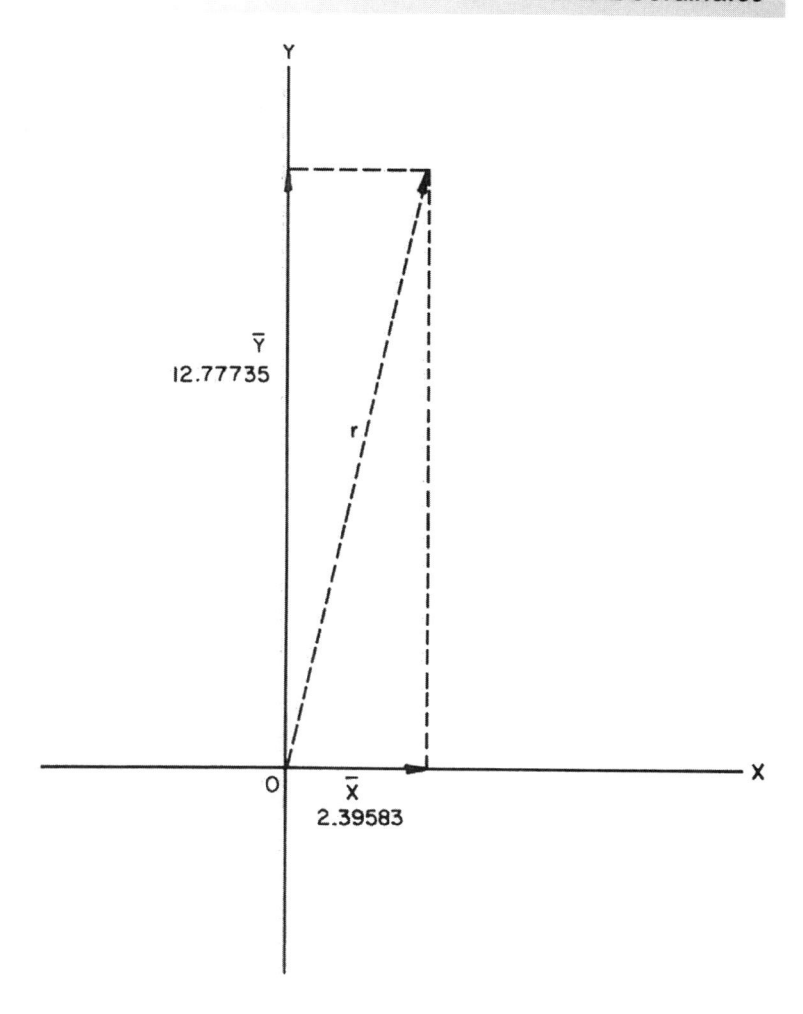

Figure 6-15.—Resultant of $\overline{X}$ and $\overline{Y}$.

If

$$r^2 = (2.39583)^2 + (12.77735)^2$$

Then

$$r = \sqrt{5.74030 + 163.25173}$$
$$= \sqrt{168.99203}$$
$$= 13$$

This is in agreement with the result found by using the method of finding the scalar resultant of two vectors.

We must now find the direction of $\overline{R}$, as follows:

Since

$$\tan \theta = \frac{y}{x}$$
$$= \frac{12.77738}{2.39583}$$
$$= 5.33480$$

Therefore

$$\theta = 79° \; 23'$$

This direction agrees with the direction found when we used the first method of finding the direction of the resultant of two vectors.

Practice Problems

Find the resultant of two vectors at right angles to each other.

1. $\overline{A} = 5\underline{/0°}$
   $\overline{B} = 10\underline{/90°}$
2. $\overline{A} = 7.5\underline{/90°}$
   $\overline{B} = 6.3\underline{/180°}$
3. $\overline{A} = 131\underline{/185°}$
   $\overline{B} = 60\underline{/275°}$
4. $\overline{A} = 65\underline{/45°}$
   $\overline{B} = 120\underline{/135°}$

Answers

1. $11.18\underline{/63° \ 26'}$
2. $9.8\underline{/130° \ 2'}$
3. $144.1\underline{/209° \ 36'}$
4. $136.5\underline{/106° \ 36'}$

Let us examine a problem of adding several vectors. We will use the method last described. The method may be used to find the addition of any number of vectors. We will consider a problem of the addition of several vectors, as follows:

EXAMPLE: Find the resultant of the vectors in figure 6-16, analytically.

SOLUTION: The vectors are given as follows:

$\overline{A}$ is $50\underline{/0°}$
$\overline{B}$ is $100\underline{/30°}$
$\overline{C}$ is $75\underline{/90°}$
$\overline{D}$ is $50\underline{/143° \ 8'}$
$\overline{E}$ is $70.7\underline{/225°}$
$\overline{F}$ is $55\underline{/315°}$

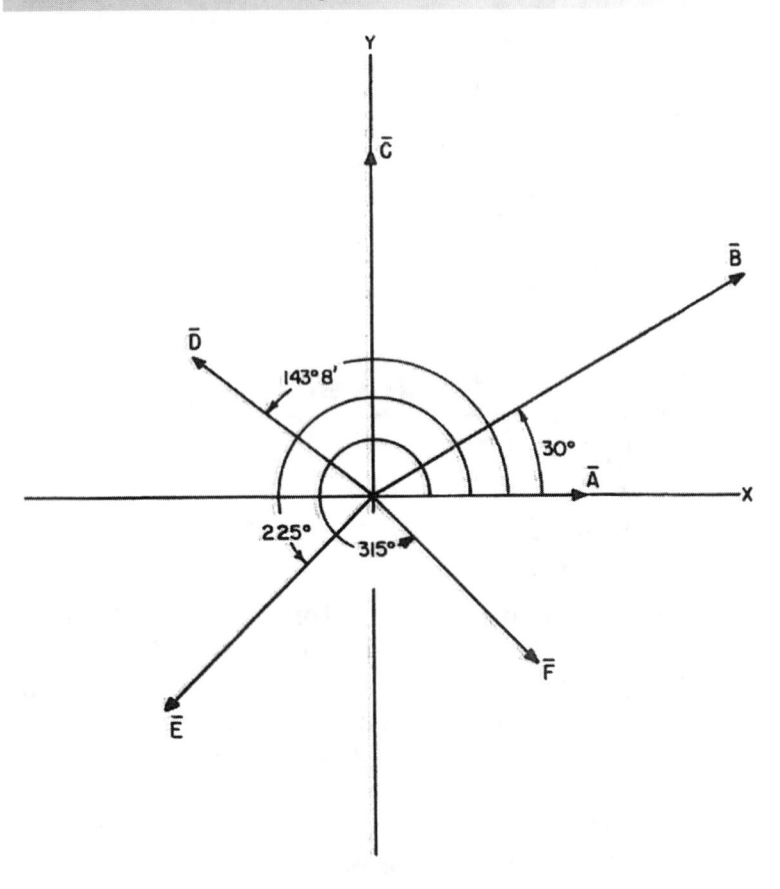

Figure 6-16.—Resultant of several vectors.

We will use the method of resolving each vector into its component X axis and Y axis coordinates. We set up the coordinate system and place each vector so that it radiates from the origin. Then we find

$$\bar{x}_a = 50 \cos 0°$$
$$= 50(1)$$
$$= 50$$

Polar Coordinates

$$\bar{y}_a = 50 \sin 0°$$
$$= 50(0)$$
$$= 0$$
$$\bar{x}_b = 100 \cos 30°$$
$$= 100(0.86603)$$
$$= 86.6$$
$$\bar{y}_b = 100 \sin 30°$$
$$= 100(0.50000)$$
$$= 50$$
$$\bar{x}_c = 75 \cos 90°$$
$$= 75(0)$$
$$= 0$$
$$\bar{y}_c = 75 \sin 90°$$
$$= 75(1)$$
$$= 75$$
$$\bar{x}_d = 50 \cos 143° \ 8'$$
$$= 50(-\cos 36° \ 52')$$
$$= 50(-0.80003)$$
$$= -40.002$$
$$\bar{y}_d = 50 \sin 143° \ 8'$$
$$= 50(\sin 36° \ 52')$$
$$= 50(0.59995)$$
$$= 30$$

$$\bar{x}_e = 70.7 \cos 225°$$
$$= 70.7 (-\cos 45°)$$
$$= 70.7 (-0.70711)$$
$$= -50$$

$$\bar{y}_e = 70.7 \sin 225°$$
$$= 70.7(-\sin 45°)$$
$$= 70.7(-0.77011)$$
$$= -50$$

$$\bar{x}_f = 55 \cos 315°$$
$$= 55 (\cos 45°)$$
$$= 55 (0.77011)$$
$$= 38.9$$

$$\bar{y}_f = 55 \sin 315°$$
$$= 55(-\sin 45°)$$
$$= 55(-0.77011)$$
$$= -38.9$$

We now collect the X axis components and the Y axis components, as follows:

Vector	$\bar{X}$	$\bar{Y}$
$\bar{A}$	50	0
$\bar{B}$	86.6	50
$\bar{C}$	0	75
$\bar{D}$	-40	30

Polar Coordinates

$\overline{E}$	- 50	- 50
$\overline{F}$	38.9	- 38.9

Adding the X axis components and the Y axis components, we find the magnitudes of $\overline{X}$ and $\overline{Y}$ as follows:

$$\overline{x}_r = 85.5$$

$$\overline{y}_r = 66.1$$

The magnitude of the resultant $\overline{R}$ is

$$r = \sqrt{(85.5)^2 + (66.1)^2}$$

$$= \sqrt{11680.67}$$

$$= 108$$

The direction is given by using the tangent function, as follows:

$$\tan \theta = \frac{66.1}{85.5}$$

$$= 0.77309$$

Therefore

$$\theta = 37° \ 42'$$

Math Made Nice-n-Easy

Multiplication

Before we discuss the mechanics of multiplication and division of vectors, in polar form, we will multiply and divide vectors in rectangular form. This will serve as an intuitive explanation of why the mechanics of polar form multiplication and division may be used.

We may express the following rectangular form vectors as complex numbers, as follows:

If
$$\overline{R}_1 = \overline{3} + \overline{4}$$
$$\overline{R}_2 = \overline{8} + \overline{5}$$

Then
$$\overline{R}_1 = 3 + 4i$$
$$\overline{R}_2 = 8 + 5i$$

And

$$\begin{aligned}
(\overline{R}_1)(\overline{R}_2) = \ & 3 + 4i \\
& \underline{8 + 5i} \\
& 24 + 32i \\
& \underline{ + 15i + 20i^2} \\
& 24 + 47i + 20i^2 \\
=\ & 24 + 47i + 20(-1) \\
=\ & 4 + 47i
\end{aligned}$$

Multiplication

Thus

$$(\overline{R}_1)(\overline{R}_2) = \overline{4} + \overline{47}$$

We now find, as shown previously, the polar form of $\overline{4} + \overline{47}$ which is as follows:

$$r = \sqrt{(4)^2 + (47)^2}$$
$$= \sqrt{2225}$$
$$= 47.2$$

And

$$\tan \theta = \frac{47}{4} = 11.75000$$
$$\theta = 85° \ 8'$$

Then

$$(\overline{R}_1)(\overline{R}_2) = 47.2 \underline{/85° \ 8'}$$

In multiplying vectors $\overline{R}_1$ and $\overline{R}_2$, in polar form, we first change to polar form as follows:

If

$$\overline{R}_1 = \overline{3} + \overline{4}$$
$$\overline{R}_2 = \overline{8} + \overline{5}$$

Then

$$\overline{R}_1 = 5\underline{/53° \ 8'}$$
$$\overline{R}_2 = 9.43\underline{/32°}$$

97

The multiplication of $\bar{R}_1$ and $\bar{R}_2$ results in a product which we will label $\bar{R}_{1,2}$. The following rule will be used:

To multiply two vectors find the product of the scalar quantities and the sum of the angles through which they have been rotated.

In our example

$$\bar{R}_1 = 5\underline{/53° \ 8'}$$

and

$$\bar{R}_2 = 9.43\underline{/32°}$$

The product of the scalar quantities is

$$(5)(9.43) = 47.2$$

and the sum of the angles is

$$(53°\underline{/8'}) + (32°) = 85° \ 8'$$

We now have the product of $\bar{R}_1$ and $\bar{R}_2$ which is $\bar{R}_{1,2}$ and is equal to $47.2\underline{/85° \ 8'}$. This result is the same as the result of multiplying the vectors in rectangular form and we intuitively understand why the mechanics of polar multiplication may be used.

Division

We will now divide vector $\bar{R}_1$ by $\bar{R}_2$ in rectangular form as follows:

If
$$\overline{R}_1 = \overline{40} + \overline{30}$$
$$\overline{R}_2 = \overline{8} + \overline{5}$$

Then
$$\overline{R}_1 = 40 + 30i$$
$$\overline{R}_2 = 8 + 5i$$

Thus
$$\left(\frac{40 + 30i}{8 + 5i}\right)\left(\frac{8 - 5i}{8 - 5i}\right)$$
$$= \frac{320 + 40i - 150i^2}{64 - 25i^2}$$
$$= \frac{470 + 40i}{89}$$
$$= \frac{470}{89} + \frac{40}{89}i$$
$$= 5.28 + 0.449i$$
$$= \overline{5.28} + \overline{0.449}$$

And
$$r = \sqrt{(5.28)^2 + (0.440)^2}$$
$$= \sqrt{27.10}$$
$$= 5.3$$

If

$$\tan \theta = \frac{0.449}{5.28}$$

$$= 0.08504$$

Then

$$\theta = 4° 52'$$

In dividing vector $\overline{R}_1$ by $\overline{R}_2$, in polar form, we first change to polar form as follows:

If

$$\overline{R}_1 = \overline{40} + \overline{30}$$

$$\overline{R}_2 = \overline{8} + \overline{5}$$

Then

$$\overline{R}_1 = 50\underline{/36° \; 52'}$$

$$\overline{R}_2 = 9.43\underline{/32°}$$

In division of vectors in polar form we will use the following rule:

To divide two vectors, in polar form, find the quotient of their scalar quantities and the difference between the angles through which they have been rotated.

Division

Thus

$$\frac{\overline{R}_1}{\overline{R}_2} = \frac{50\underline{/36°\ 52'}}{9.43\underline{/32°}}$$

$$= 5.3\underline{/4°\ 52'}$$

This result is the same as the result obtained by dividing vector $\overline{R}_1$ by $\overline{R}_2$, in rectangular form, and we intuitively understand why the mechanics of polar division may be used.

Practice Problems

Multiply the following vectors:

1. $(5\underline{/10°})\ (10\underline{/5°})$
2. $(8.3\underline{/6°})\ (1.1\underline{/73°})$
3. $(6.2\underline{/52°})\ (8\underline{/200°})$
4. $(100\underline{/45°})\ (30\underline{/20°})$

Answers

1. $50\underline{/15°}$
2. $9.13\underline{/79°}$
3. $49.6\underline{/252°}$
4. $3000\underline{/65°}$

Practice Problems

Perform the indicated division:

1. $\dfrac{64/24°}{8/24°}$

2. $\dfrac{300/24°}{20/8°}$

3. $\dfrac{620/154°}{5/142°}$

4. $\dfrac{64/18°}{16/27°}$

Answers

1. $8/0°$
2. $15/16°$
3. $124/12°$
4. $4/-9°$

It follows that a vector can be raised to any integral or fractional power. To square a vector, square the scalar quantity and multiply the angle by 2.

EXAMPLE: Square the vector $8/32°$

Division

SOLUTION: $(8\underline{/32°})^2$

$= (8)^2\underline{/32°\ (2)}$

$= 64\underline{/64°}$

To cube a vector, cube the scalar quantity and multiply the angle by 3.

EXAMPLE: Cube the vector $3\underline{/4°}$

SOLUTION: $(3\underline{/4°})^3$

$= (3)^3\underline{/4°\ (3)}$

$= 27\underline{/12°}$

To find the square root of a vector, extract the square root of the scalar quantity and divide the angle by 2.

EXAMPLE: Find the square root of the vector $16\underline{/70°}$.

SOLUTION: $\sqrt{16\underline{/70°}}$ or $(16\underline{/70°})^{1/2}$

$= \sqrt{16}\underline{/70°} \div 2$

$= 4\underline{/35°}$

To find the cube root of a vector, extract the cube root of the scalar quantity and divide the angle by 3.

EXAMPLE: Find the cube root of the vector $27\underline{/33°}$.

SOLUTION: $\sqrt[3]{27\underline{/33°}}$ or $(27\underline{/33°})^{1/3}$

$= \sqrt[3]{27}\underline{/33° \div 3}$

$= 3\underline{/11°}$

Practice Problems

Perform the indicated operations:

1. $(10\underline{/20°})^2$
2. $(4\underline{/10°})^3$
3. $(64\underline{/90°})^{1/2}$
4. $(64\underline{/90°})^{1/3}$

Answers

1. $100\underline{/40°}$
2. $64\underline{/30°}$
3. $8\underline{/45°}$
4. $4\underline{/30°}$

Chapter 7
Applications of Vectors

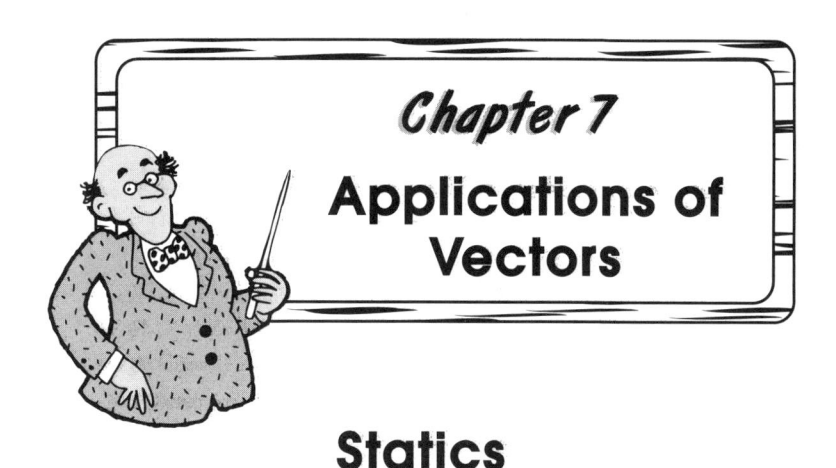

Statics

Statics is a branch of physics that deals with bodies at rest. In this chapter we will make use of the previous investigation of vectors to establish the mathematical basis necessary for an understanding of static equilibrium. Since forces acting upon bodies have magnitude and direction, they may be represented by vectors.

Definitions and Terms

The following paragraphs include definitions and terms which will be used in this chapter. The definitions used will clarify the meanings of the discussions on static equilibrium.

Equilibrium

If a body undergoes no change in its motion, it is said to be in a state of equilibrium. We will discuss a body at rest as indicated by the term static equilibrium. Balanced forces may act upon a body in static equilibrium, but no motion, neither translatory nor rotary, will occur. In order for bodies to be in static equilibrium, two conditions are required. These two conditions are (1) the body must not have translatory motion, and (2) the body must not have rotary motion.

Translation

Translation, as defined, is motion independent of rotation. Attention is called to the fact that translation involves magnitude and direction of motion and hence can be described by vectors.

Rotation

Rotation, as defined, is the turning motion of a body, such as a wheel turning. Rotation is independent of translation.

Translational Equilibrium

If a body at rest is acted upon by an unbalanced force, it will be set into motion. This motion is called translation. All particles of the moving body will have at any instant the same velocity and direction of motion. Two forces acting in the same line upon a body must be equal in magnitude, but opposite in direction, if the body is to remain in equilibrium. We will consider our translations to be confined to the XY plane.

First Condition

The first condition of equilibrium may be stated as follows: For a particle to be in equilibrium the sum of the vectors (forces) acting in any direction upon that particle must equal zero.

In figure 7-1, an iron block is resting on a table. The weight of the block is directed downward; thus the table must exert a force equal and opposite to the weight of this iron block. The block has weight, W, and the table exerts a push, P, upward against the block. Since the bodies are in equilibrium, there can be no unbalanced force. Thus,

$$W = P \text{ and } \overline{W} = \overline{P}$$

The weight of the block can be represented by a vector because we know the force and direction exerted by the weight. The magnitude and direction of the push (P) by the table is also known, and it can be represented by a vector. The vectors $\overline{W}$ and $\overline{P}$ are shown in figure 7-1. The weight is in equilibrium because the sum of the vectors acting upon it is equal to zero. We call these two forces parallel concurrent forces. We will also call $\overline{P}$ the equilibrant of $\overline{W}$. It is relatively easy to find the equilibrant of two or more vectors which are acting upon the same point. We first find the resultant of the vectors. The equilibrant of the resultant will have the same magnitude but will be opposite in direction. In figure 7-2 the resultant of $\overline{A}$ and $\overline{B}$ is $\overline{R}$. The magnitude of $\overline{R}$

is 18 and the direction is 56°18'. The magnitude of $\overline{C}$, the equilibrant of $\overline{R}$, is 18 and the direction is 236°18'. The sum of $\overline{R}$ and $\overline{C}$ is zero; therefore, a point 0 is in equilibrium.

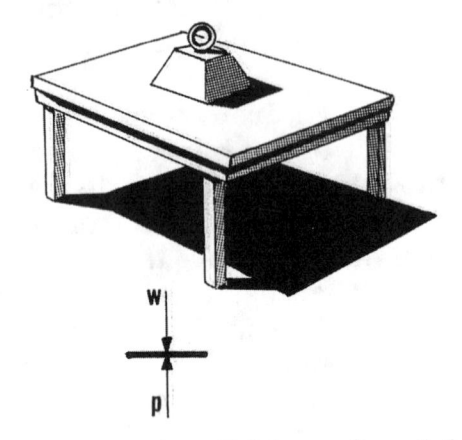

Figure 7-1.—Table and weight.

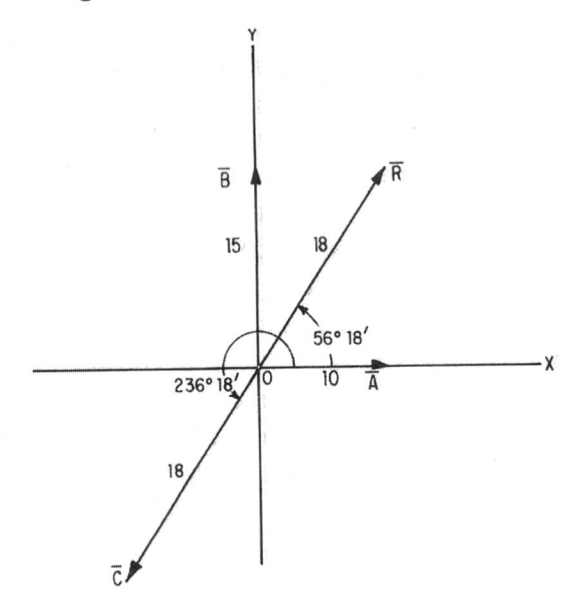

Figure 7-2.—The equilibrant.

First Condition

In figure 7-2, the vectors $\overline{A}$ and $\overline{B}$ are called nonparallel concurrent vectors. All vectors can be resolved into horizontal and vertical components. Since the sum of all forces acting on a particle must be equal to zero, to satisfy the condition of equilibrium, we can say that the sum of all vertical components must equal zero, and the sum of all the horizontal components must equal zero. The symbols for this condition of equilibrium are:

$$\Sigma \overline{X} = 0$$

and

$$\Sigma \overline{Y} = 0$$

The symbol Σ is the Greek letter, sigma, and means "the sum of." Thus, $\Sigma \overline{X}$ equals 0 means that the sum of the vectors along the X axis equals zero.

We may show graphically that a particle P is in equilibrium while being acted upon by $\overline{A}$, $\overline{B}$, $\overline{C}$, $\overline{D}$, and $\overline{E}$, as in figure 7-3, by drawing a polygon of forces. If the polygon of forces is closed, there is no resultant force acting upon particle $\overline{P}$, and that particle is in equilibrium.

We will now examine the condition of equilibrium of a point which is acted upon by nonparallel concurrent forces.

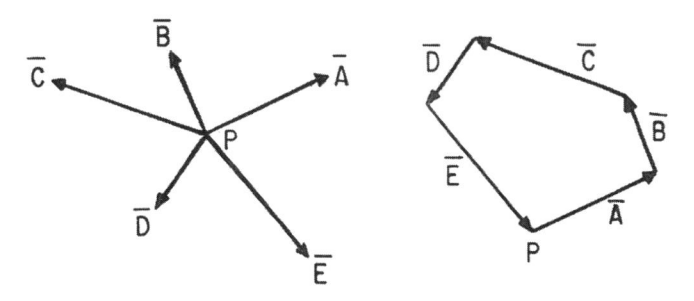

Figure 7-3.—Polygon of forces.

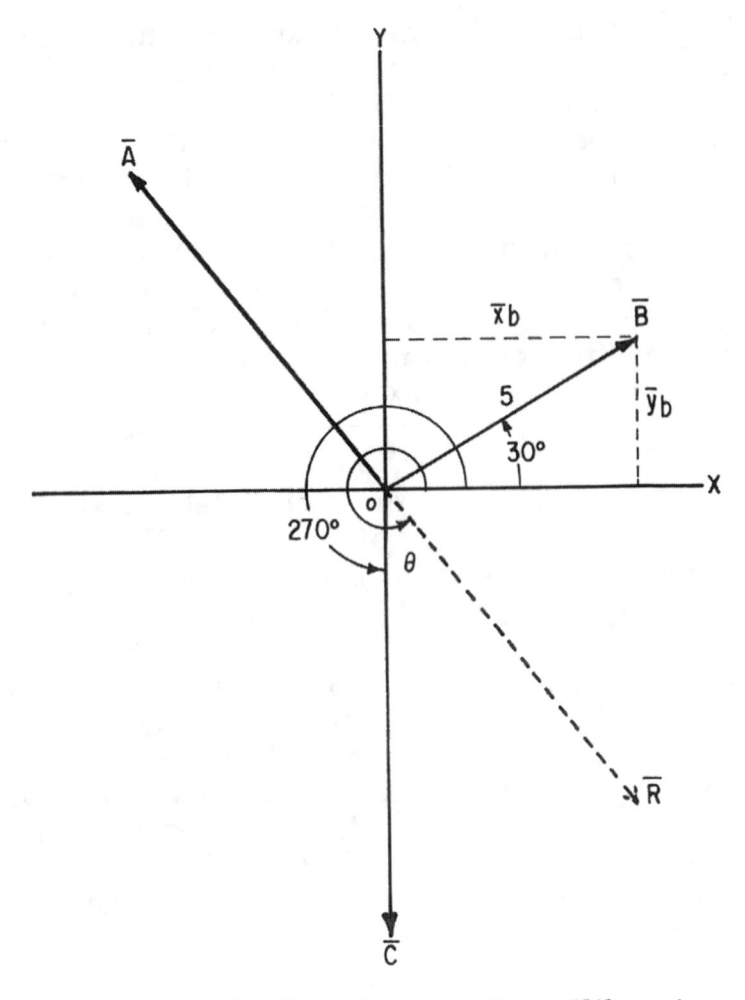

Figure 7-4.—Resultant and equilibrant.

EXAMPLE: We are to find the force of $\overline{A}$, in figure 7-4, in order that point 0 will remain in equilibrium while being acted upon by $\overline{B}$ and $\overline{C}$.

SOLUTION: We are looking for the equilibrant of the resultant of $\overline{B}$ and $\overline{C}$. The resultant of $\overline{B}$ and $\overline{C}$, called $\overline{R}$, is found as follows:

First Condition

$$\bar{y}_b = 5 \sin 30°$$
$$= 5(0.50000)$$
$$= 2.5$$

$$\bar{x}_b = 5 \cos 30°$$
$$= 5(0.86603)$$
$$= 4.3$$

$$\bar{y}_c = 8 \sin 270°$$
$$= 8(-1.00000)$$
$$= -8$$

$$\bar{x}_c = 8 \cos 270°$$
$$= 8(0.00000)$$
$$= 0$$

We now add the X axis components and the Y axis components and find that

$$\bar{Y} = 2.5 + (-8)$$
$$= -5.5$$

$$\bar{X} = 4.3 + 0$$
$$= 4.3$$

and

$$\tan \theta = \frac{-5.5}{4.3}$$

$$= -1.27907$$

thus

$$\theta = 308° 1'$$

and

$$r = \sqrt{(4.3)^2 + (-5.5)^2}$$

$$= \sqrt{48.74}$$

$$= 6.9$$

therefore our resultant is

$$6.9 \underline{/308° 1'}$$

and the equilibrant is

$$6.9 \underline{/128° 1'}$$

In some cases we are given the vectors by the problem and can easily find our solution.

EXAMPLE: A boy in a swing, as shown in figure 7-5, weighs 70 pounds and is pulled backward with a force of 30 pounds. Find the force the ropes exert on the swing; also find the angle the ropes make with the horizontal axis.

First Condition

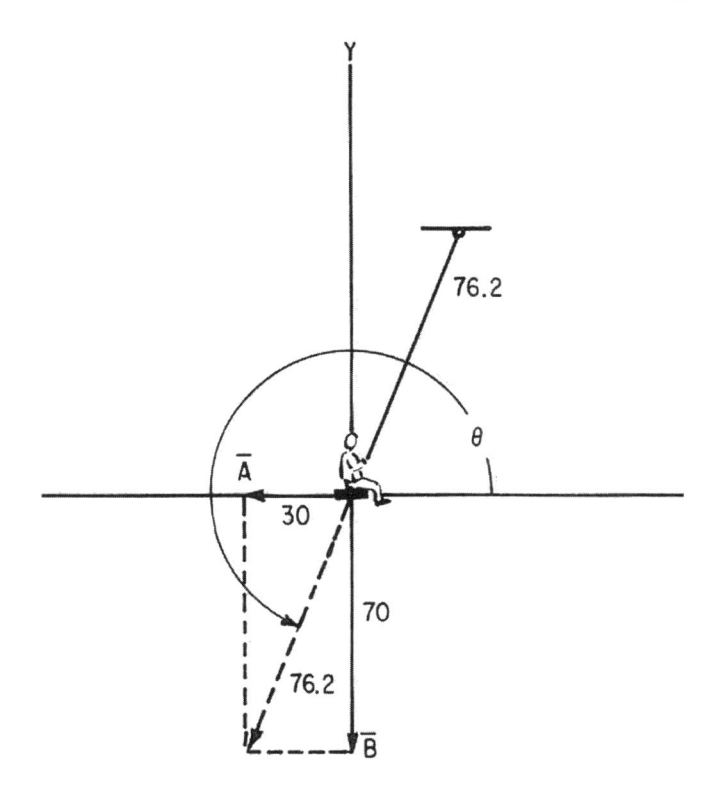

Figure 7-5.—Boy in a swing.

SOLUTION: We must find the resultant of the two vectors given and then find the equilibrant of the resultant. This is done as follows:

$$\bar{x}_a = -30$$
$$\bar{y}_a = 0$$
$$\bar{x}_b = 0$$
$$\bar{y}_b = -70$$

and

thus
$$\tan\theta = \frac{-70}{-30} = 2.33333$$

and
$$\theta = 246° \ 48'$$

$$r = \sqrt{(-30)^2 + (-70)^2}$$
$$= \sqrt{5800}$$
$$= 76.2$$

therefore the resultant is

$$76.2 / 246° \ 48'$$

and the equilibrant is

$$76.2 / 66° \ 48'$$

Thus the ropes exert a 76.2-pound force at 66°48' on the swing.

Practice Problems

Find the equilibrant of the following vectors.

1. $35/0°$ and $60/90°$
2. $7/35°$ and $9/125°$
3. $12/15°$ and $7/25°$
4. $9/55°$ and $10/100°$

Answers

1. $69.4/239°\ 45'$
2. $11.4/267°\ 7'$
3. $18.7/198°\ 43'$
4. $17.5/257°\ 16'$

Free Body Diagrams

One of the distinct advantages of vectors is that a vector may be substituted for the cable or member of a mechanism it is going to represent. As seen in figure 7-6 vectors may be substituted for the cables holding the weight W. Starting at point O, a vector representing the tension in cable MO can be drawn, and vectors may also be drawn for the tension in cables NO and OW. This will give us the forces acting on particle O. Figure 7-6 (B) is called a free body diagram, and vector $\overline{A}$ represents the tension in cable MO. Vectors $\overline{B}$ and $\overline{W}$ represent the tensions in NO and WO, respectively.

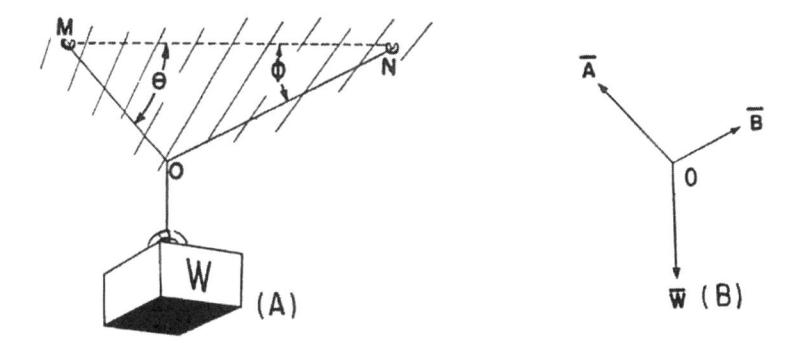

Figure 7-6.—Single weight.

Math Made Nice-n-Easy

Free body diagrams are very important in mechanics and the student should learn to draw these diagrams with ease. In a free body diagram, a member of a mechanism is replaced by a vector representing the force in that member and acting in the same direction as the member. The student should pay particular attention to the magnitude of the vector which represents a member of the mechanism. In figure 7-6, notice that the vector representation for MO is longer and the vector representation for NO is shorter in the free body diagram (B) than they appear in the pictorial view (A).

We may use the free body diagram to graphically verify our mathematical solution to a problem. (Refer again to fig. 7-5.) We find the boy in the swing to be in equilibrium and we will use a free body diagram to verify this. We draw our diagram as shown in figure 7-7 (A) where vector $\overline{C}$ is the equilibrant of the resultant.

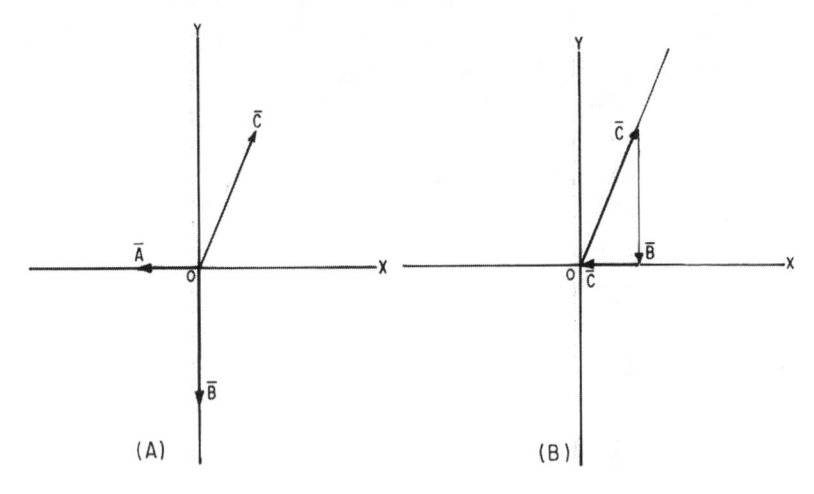

Figure 7-7.—Closed loop.

Free Body Diagrams

We draw the vectors $\overline{A}$, $\overline{B}$, and $\overline{C}$, initial point to terminal point, as shown in figure 7-7 (B). If the vectors form a closed loop, we have the sum of the vectors equal to zero and have present a state of equilibrium.

We have discussed parallel concurrent forces and nonparallel concurrent forces. In the following paragraphs we will discuss noncurrent parallel forces, remembering that we are still under the requirements for the first condition of equilibrium.

In figure 7-8 (A), we find a board balanced on and supported by a fulcrum. Draw the free body diagram as shown in figure 7-8 (B). Assume the board weighs so little that it is insignificant. Consider forces in a downward direction to be negative (-) and those upward to be positive (+). For equlibrium, we must have ΣX equal to 0 and ΣY equal to 0. We have no X axis components, therefore ΣX equals 0. The ΣY equals 0 because we have a state of equilibrium. Therefore

and
$$-\overline{A} - \overline{B} + \overline{C} = 0$$
$$\overline{A} = 42 \text{ lbs}$$

therefore
$$\overline{B} = 18 \text{ lbs}$$
$$\overline{C} = 60 \text{ lbs}$$

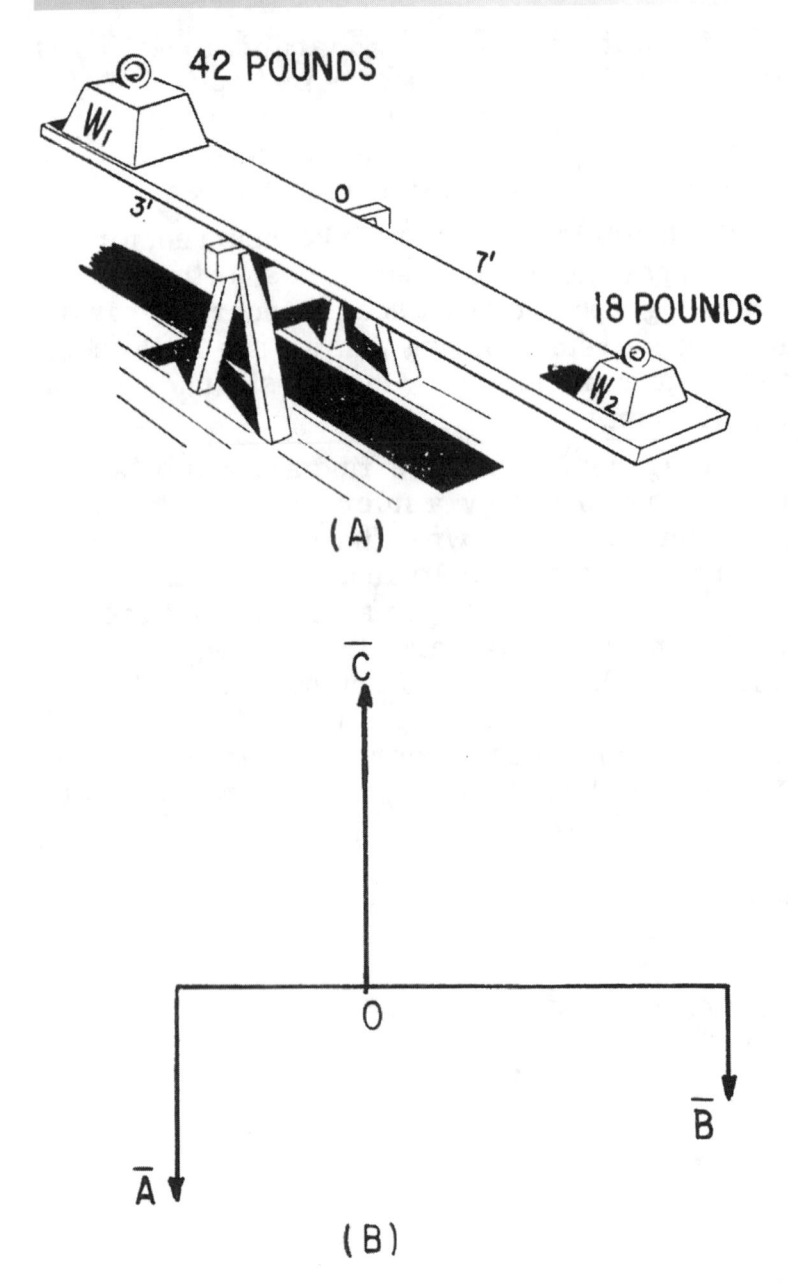

Figure 7-8.—Parallel nonconcurrent forces.

Free Body Diagrams

Practice Problems

Draw the free body diagrams for the member indicated in the following figures (emphasize direction and not magnitude):

1. Figure 7-9 (A) (Bar)
2. Figure 7-9 (B) (Boom)
3. Figure 7-9 (C) (Bar)
4. Figure 7-9 (D) (Point O)

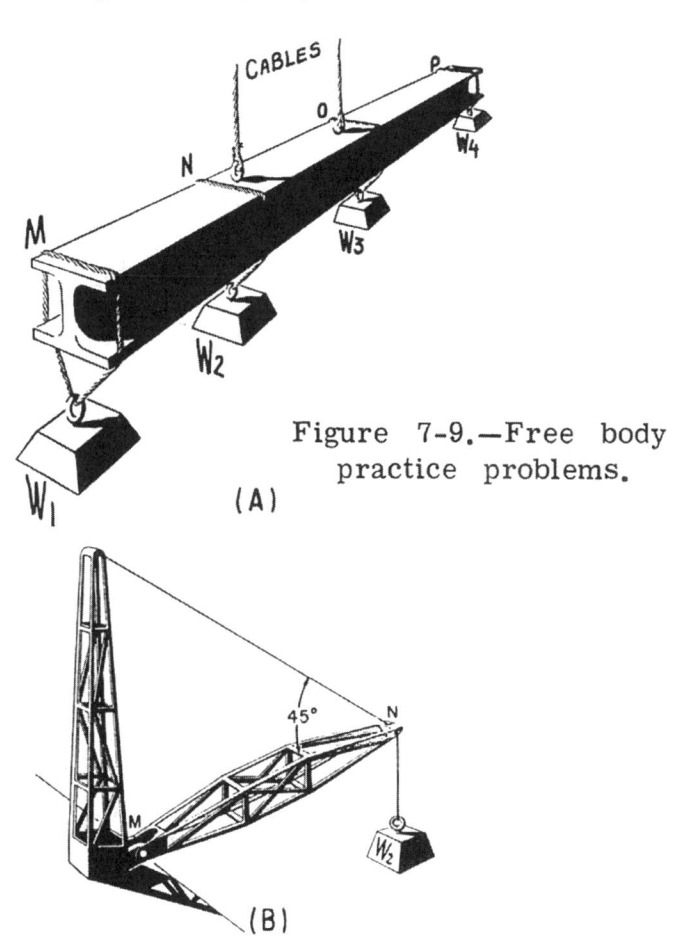

Figure 7-9.—Free body practice problems.

119

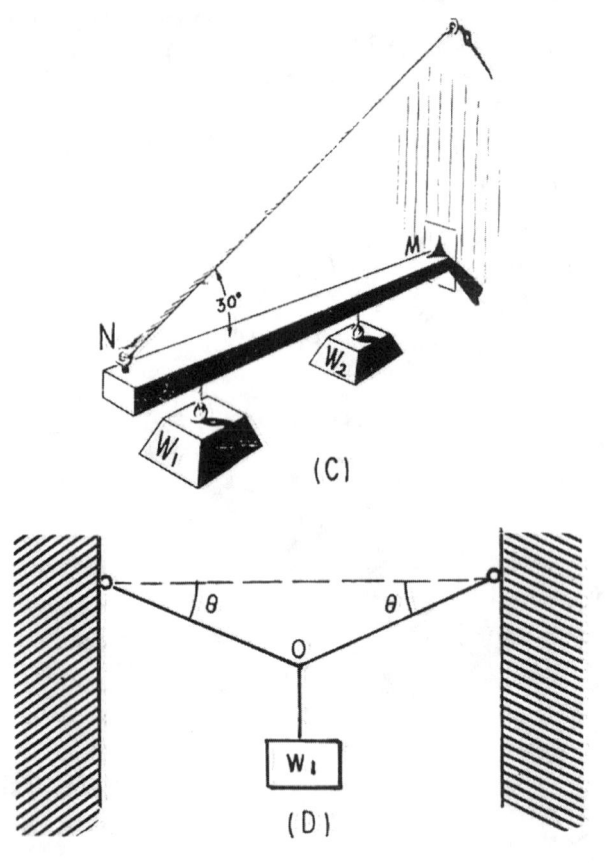

Figure 7-9.—Free body practice problems. *(continued)*

Answers

1. Figure 7-10 (A)
2. Figure 7-10 (B)
3. Figure 7-10 (C)
4. Figure 7-10 (D)

Free Body Diagrams

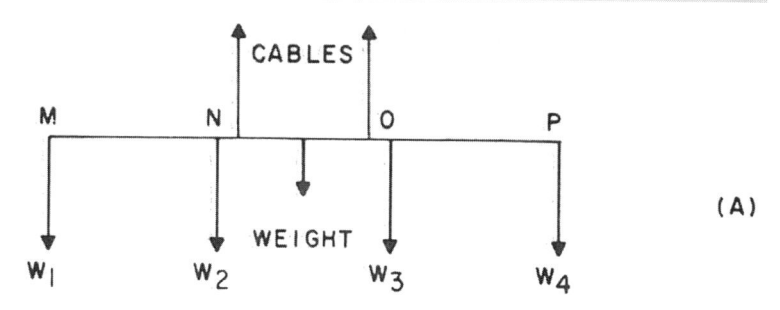

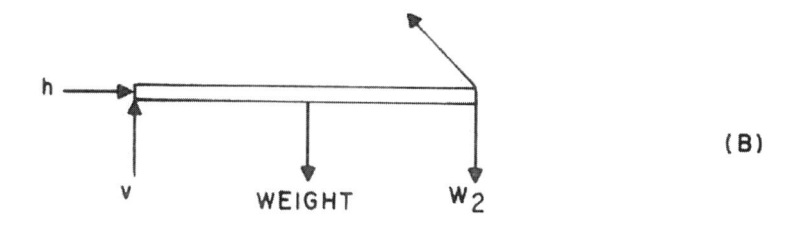

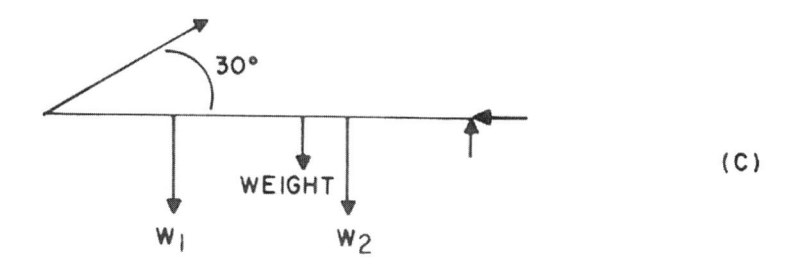

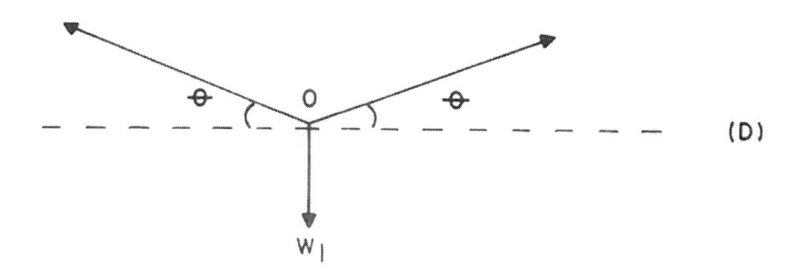

Figure 7-10.—Free body answers.

Rotational Equilibrium

The first condition of equilibrium guaranteed that there would be only translatory motion. It was stated that there was a distinction between the motion of translation and rotation. The second condition of equilibrium concerns the forces tending to rotate a body.

Second Condition

Figure 7-11 shows a body acted upon by two equal and opposite forces, F_1 and F_2. The sum of the forces in the horizontal direction equals zero, and there is no translatory motion. It is clear that there will be rotation of the body. These two equal and opposite forces not acting along the same line constitute a couple and cause a moment to be produced. The term couple is defined as two equal forces acting on a body but in opposite directions and not along the same line. For a body acted upon by a couple to remain in equilibrium, it must be acted upon by another couple equal in magnitude but opposite in direction.

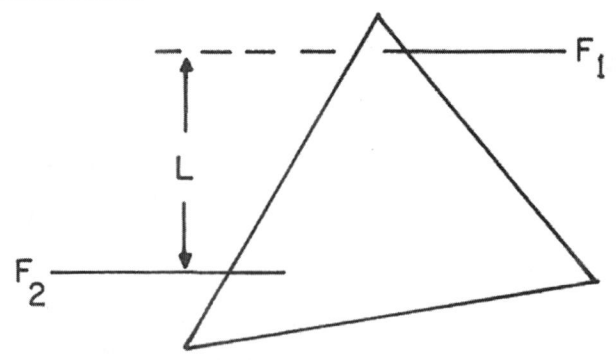

Figure 7-11.—Forces forming a couple.

Second Condition

The magnitude of a couple is the perpendicular distance between the forces multiplied by one of the forces. This product is called the moment of the couple. We will use M to indicate a moment and we can say, to fulfill the conditions of equilibrium, that ΣM equals 0. That is, the sum of all the moments acting upon a body must equal zero to maintain equilibrium. Clockwise moments, such as in figure 7-11, are positive and counterclockwise moments are negative. Our statement that the sum of all the moments acting upon a body must be zero, that is, ΣM equals 0, is called the second condition for equilibrium.

Assume that the body shown in figure 7-11 will rotate about a point halfway between the two forces. A moment, defined as a force acting on a lever arm L, is present for both of the forces. The moment acting on the body in figure 7-11 will be

$$F_1 \left(\frac{L}{2}\right) + F_2 \left(\frac{L}{2}\right)$$

Since
$$F_1 = F_2$$

then
$$M = F_1 \left(\frac{L}{2}\right) + F_1 \left(\frac{L}{2}\right)$$
$$= F_1 L$$

If it were assumed that the body were to rotate about the point upon which F_2 acted, then the lever arm would be L for F_1 and zero for F_2. And again M equals $F_1 L$. Hence, the moment of a couple is one of the forces multiplied by the distance between them. The definition for moment of a couple holds. The dimensions of a moment will include a distance as well as a force. The effect of a force upon the rotation is the perpendicular distance from the rotation point to the line of action of the force. In the English system the most used term is foot-pounds.

EXAMPLE: Calculate the moment of a couple consisting of two forces, F_1 (equal to 20 pounds) and F_2 (equal to 20 pounds), acting directly opposite to each other at a distance of 3 feet. The moment of this couple is M equals FL or

$$M = (20)(3) = 60 \text{ ft-lb}$$

Notice that in this example there is no balance of moments; that is, ΣM does not equal 0, and the conditions for equilibrium are not met.

We now put to use the first and second conditions for equilibrium. That is,

$$\Sigma Y = 0$$

$$\Sigma X = 0$$

$$\Sigma M = 0$$

In figure 7-12 (A) we have a board balanced on a fulcrum and we are to find the weight W_2 and the force F on the fulcrum if W_1 equals 12 pounds and the distances are as shown. There are no horizontal forces; therefore ΣX equals 0. We draw the free body diagram as shown in figure 7-12 (B) and find the solution as follows:

Second Condition

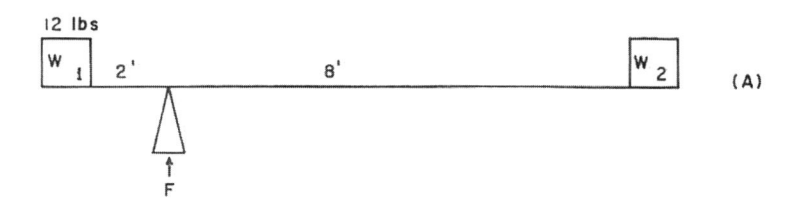

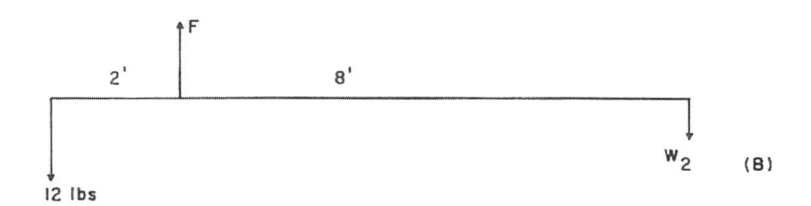

Figure 7-12.—Forces and moments.

If

$$W_1 = 12 \text{ lb}$$

$$L_1 = 2 \text{ ft}$$

$$W_2 = \text{unknown}$$

$$L_2 = 8 \text{ ft}$$

$$F = \text{unknown}$$

then

$$L_1 W_1 = L_2 W_2$$
$$= (12 \text{ lb})(2 \text{ ft})$$
$$= 24 \text{ ft-lb}$$

Therefore

$$L_2 W_2 = 24 \text{ ft-lb}$$
$$(8 \text{ ft})(W_2) = 24 \text{ ft-lb}$$
$$W_2 = 3 \text{ lb}$$

Thus

$$\Sigma M = 0$$

and

$$W_1 + W_2 - F = 0$$
$$F = 12 \text{ lb} + 3 \text{ lb}$$
$$= 15 \text{ lb}$$

Thus

$$\Sigma Y = 0$$

A very useful theorem that originates from the second condition of equilibrium states that: If three nonparallel forces acting upon a body produce equilibrium, their lines of action must pass through a common point. In other words, the three conditions ΣX equals 0, ΣY equals 0, and ΣM equals 0 must be satisfied for equilibrium; and, in order for three nonparallel forces to produce zero moment, the lines of action of the forces must pass through a common point, thus having zero lever arm.

Center of Gravity

The earth's gravitational field attracts each particle in a body and the weight of that body is regarded as a system of parallel forces acting upon each particle of the body. All of these parallel forces can be replaced by a single force equal to their sum. The point of application of this single force is called the center of gravity (or C. G.) of the body. For bodies of simple shape and uniform density, the C. G. is at the geometric center and can be found by inspection.

The C. G. of an irregularly shaped body can be found by suspending the body from three different points on the body. In each case the body will come to rest (equilibrium) with its C. G. directly beneath the point of suspension. The intersection of any two of these lines will determine the C. G. The third line should also pass through this intersection and thus may be used to check the result. Figure 7-13 (A) shows this simple system for finding the C. G. Figure 7-13 (B) shows that in some cases the C. G. may fall outside of the body.

(A)   (B)

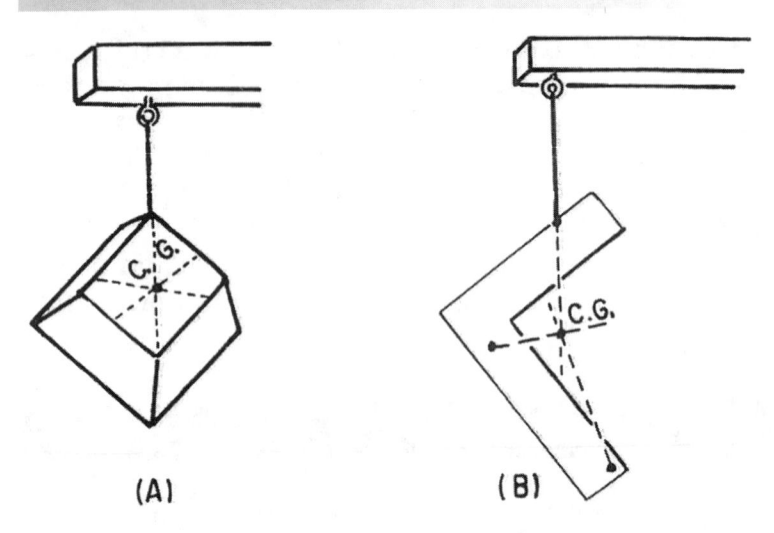

Figure 7-13.—Center of gravity.

Applications

The greatest difficulty encountered in solving problems dealing with equilibrium is finding all of the forces acting upon a body. The use of a free body diagram will aid in eliminating this difficulty.

The procedure recommended for solving static equilibrium problems is as follows:

1. Sketch the system, taking into account all known facts, and assign symbols to all of the knowns and unknowns.

2. Select a member that involves one or more of the unknowns and construct a free body diagram.

3. Write the equations obtained from ΣX equals 0, ΣY equals 0, and ΣM equals 0.

4. Solve these equations for the unknowns.

Applications

5. Continue the process from one side of a structure to the other side.

EXAMPLE: Consider the ladder standing against a building in figure 7-14 (A) and making an angle of 60° with the ground. The ladder is 16 feet long and weighs 50 pounds.

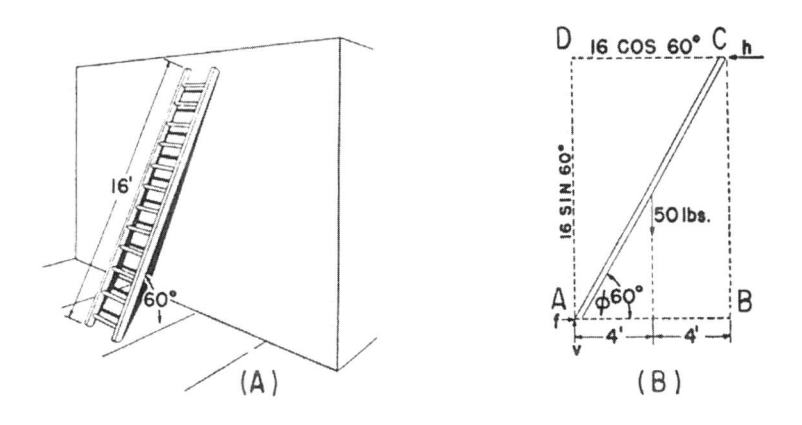

Figure 7-14.—Ladder problem.

SOLUTION: We sketch the free body diagram as shown in figure 7-14 (B) and assign symbols to the known and unknown forces. The arrows indicate the directions of the forces and h and v represent horizontal and vertical components of a force. The frictional force f holds the ladder from slipping, h is the horizontal force of the wall pushing against the ladder, and v is the vertical force which the ground exerts on the ladder. We assume all the weight of the ladder to be located at the center of gravity and assign the letter W to indicate this weight. The ladder is in a state of equilibrium and we have the following:

$$\Sigma X = 0$$

$$\Sigma Y = 0$$

$$\Sigma M = 0$$

therefore

$$f - h = 0$$

$$W - v = 0$$

We use trigonometry to find $\overline{AB}$ and $\overline{BC}$, as follows:

$$\overline{AB} = r \cos \phi$$
$$= 16 \cos 60°$$
$$= 16 (0.50000)$$
$$= 8$$

and

$$\overline{BC} = r \sin \phi$$
$$= 16 \sin 60°$$
$$= 16 (0.86603)$$
$$= 13.85$$

Using similar triangles we find that W is located directly above the midpoint of $\overline{AB}$.

Next, take the moments about the bottom of the ladder and in that way the two forces (f and v) have zero lever arm and are eliminated. The moments (clockwise) are as follows:

$$\Sigma M = 0$$

$$= 4W - h(16 \sin 60°)$$

Notice that we used the perpendicular distances from point A to where the forces were applied.

We now have equations as follows:

$$f - h = 0$$

$$W - v = 0$$

$$4W - h (16 \sin 60°) = 0$$

and substituting known values, then solving, we find

$$W - v = 0$$

$$50 \text{ lb} - v = 0$$

$$v = 50 \text{ lb}$$

and

$$4W - h (16 \sin 60°) = 0$$

$$4 (50 \text{ lb}) - h (13.85) = 0$$

$$h = 14.4 \text{ lb}$$

and

$$f - h = 0$$

$$f - 14.4 = 0$$

$$f = 14.4 \text{ lb}$$

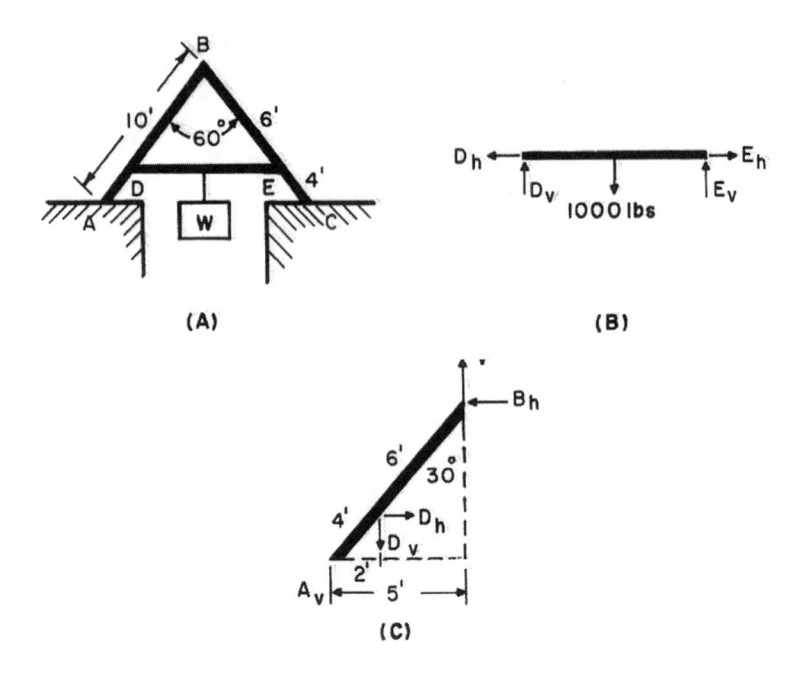

Figure 7-15.—A-type frame.

EXAMPLE: Determine the forces acting upon the members of the A-type frame as shown in figure 7-15 (A). The horizontal surface is considered smooth and no horizontal force can be exerted on the legs of the frame. A weight of 1,000 pounds hangs from the crossbar. The frame is considered as having no weight.

Applications

SOLUTION: Draw the free body diagrams and assign symbols as shown in figures 7-15(B) and (C). Since the system is symmetrical, the reactions at A and C are equal, and each is equal to 500 pounds (each carries half the load). Thus, A_V equals 500 pounds and C_V equals 500 pounds. The forces D_V and E_V can be found from the diagram of the crossbar in figure 7-15 (B) by taking ΣM about D.
Thus

$$\Sigma M = 3W - E_V(6) = 0$$

$$= 3(1000) - E_V(6)$$

$$E_V = 500 \text{ lb}$$

and from symmetry

$$D_V = 500 \text{ lb}$$

These forces E_V and D_V are upward to oppose the weight on the bar; thus this member must exert the same forces downward on the inclined member.

It is apparent that BC pushes upward against AB. This force is unknown, but it does have a vertical and horizontal component.

Using the two conditions of equilibrium we find the following:

$$\Sigma \overline{X} = D_h - B_h = 0$$
$$\Sigma \overline{Y} = A_V + B_V - D_V = 0$$
$$\Sigma M = 5A_V - 3D_V - (6 \sin 60°)D_h = 0$$

Thus

$$\Sigma \overline{Y} = 500 + B_v - 500 = 0$$

$$B_v = 0$$

$$\Sigma M_b = 5(500 \text{ lb}) - 3(500 \text{ lb}) - 6(0.86603) = 0$$

$$D_h = 192 \text{ pounds}$$

$$\Sigma \overline{X} = 192 \text{ lb} - B_h = 0$$

$$B_h = 192 \text{ lb}$$

From symmetry we find the following:

$$E_h = D_h = 192 \text{ pounds}$$

$$E_v = D_v = 500 \text{ pounds}$$

The magnitude and direction of the X axis and Y axis forces may be used to find the forces and their directions.

One important thing to remember when taking ΣM equal to 0 is to take the moments about some point that will eliminate one or more of the unknowns. In the last example, the moment equation was taken about point B and eliminated the forces B_h and B_v. The moment equations can be taken about any point and more than one moment equation can be taken, if necessary.

Applications

Practice Problems

Find the required information in the following:

1. Two boys pull a wagon, each with a force of 35 pounds. The angle between the ropes on which the boys are pulling is 30°. What is the resultant pull on the wagon?

2. A large portrait weighs 100 pounds, and is supported by a wire 10 feet long which is hooked to the picture at two points 5 feet apart. Find the tension in the wire.

3. A 180-pound man is standing half-way up a 20-foot, 20-pound ladder. The bottom of the ladder is 4 feet from the base of the vertical wall it is leaning against. Find the forces exerted on the ladder. (Use same symbols as shown in fig. 7-14 (B).)

4. A bar of uniform weight, 12 feet long and weighing 7 pounds, is supported by a fulcrum which is 4 feet from the left end. If a 10-pound weight is hung from the left end, find the weight needed at the right end to hold the bar in equilibrium and find the force with which the fulcrum pushes against the bar.

Answers

1. 67.6 pounds
2. 57.8 pounds
3. f = 20.4 pounds, h = 20.4 pounds, v = 200 pounds
4. 3.25 pounds and 20.25 pounds

Common Logarithms of Numbers

No.	0	1	2	3	4	5	6	7	8	9
10	0000	0043	0086	0128	0170	0212	0253	0294	0334	0374
11	0414	0453	0492	0531	0569	0607	0645	0682	0719	0755
12	0792	0828	0864	0899	0934	0969	1004	1038	1072	1106
13	1139	1173	1206	1239	1271	1303	1335	1367	1399	1430
14	1461	1492	1523	1553	1584	1614	1644	1673	1703	1732
15	1761	1790	1818	1847	1875	1903	1931	1959	1987	2014
16	2041	2068	2095	2122	2148	2175	2201	2227	2253	2279
17	2304	2330	2355	2380	2405	2430	2455	2480	2504	2529
18	2553	2577	2601	2625	2648	2672	2695	2718	2742	2765
19	2788	2810	2833	2856	2878	2900	2923	2945	2967	2989
20	3010	3032	3054	3075	3096	3118	3139	3160	3181	3201
21	3222	3243	3263	3284	3304	3324	3345	3365	3385	3404
22	3424	3444	3464	3483	3502	3522	3541	3560	3579	3598
23	3617	3636	3655	3674	3692	3711	3729	3747	3766	3784
24	3802	3820	3838	3856	3874	3892	3909	3927	3945	3962
25	3979	3997	4014	4031	4048	4065	4082	4099	4116	4133
26	4150	4166	4183	4200	4216	4232	4249	4265	4281	4298
27	4314	4330	4346	4362	4378	4393	4409	4425	4440	4456
28	4472	4487	4502	4518	4533	4548	4564	4579	4594	4609
29	4624	4639	4654	4669	4683	4698	4713	4728	4742	4757
30	4771	4786	4800	4814	4829	4843	4857	4871	4886	4900
31	4914	4928	4942	4955	4969	4983	4997	5011	5024	5038
32	5051	5065	5079	5092	5105	5119	5132	5145	5159	5172
33	5185	5198	5211	5224	5237	5250	5263	5276	5289	5302
34	5315	5328	5340	5353	5366	5378	5391	5403	5416	5428
35	5441	5453	5465	5478	5490	5502	5514	5527	5539	5551
36	5563	5575	5587	5599	5611	5623	5635	5647	5658	5670
37	5682	5694	5705	5717	5729	5740	5752	5763	5775	5786
38	5798	5809	5821	5832	5843	5855	5866	5877	5888	5899
39	5911	5922	5933	5944	5955	5966	5977	5988	5999	6010
40	6021	6031	6042	6053	6064	6075	6085	6096	6107	6117
41	6128	6138	6149	6160	6170	6180	6191	6201	6212	6222
42	6232	6243	6253	6263	6274	6284	6294	6304	6314	6325
43	6335	6345	6355	6365	6375	6385	6395	6405	6415	6425
44	6435	6444	6454	6464	6474	6484	6493	6503	6513	6522
45	6532	6542	6551	6561	6571	6580	6590	6599	6609	6618
46	6628	6637	6646	6656	6665	6675	6684	6693	6702	6712
47	6721	6730	6739	6749	6758	6767	6776	6785	6794	6803
48	6812	6821	6830	6839	6848	6857	6866	6875	6884	6893
49	6902	6911	6920	6928	6937	6946	6955	6964	6972	6981
50	6990	6998	7007	7016	7024	7033	7042	7050	7059	7067
51	7076	7084	7093	7101	7110	7118	7126	7135	7143	7152
52	7160	7168	7177	7185	7193	7202	7210	7218	7226	7235
53	7243	7251	7259	7267	7275	7284	7292	7300	7308	7316
54	7324	7332	7340	7348	7356	7364	7372	7380	7388	7396
No.	0	1	2	3	4	5	6	7	8	9

Common Logarithms of Numbers (cont'd)

No.	0	1	2	3	4	5	6	7	8	9
55	7404	7412	7419	7427	7435	7443	7451	7459	7466	7474
56	7482	7490	7497	7505	7513	7520	7528	7536	7543	7551
57	7559	7566	7574	7582	7589	7597	7604	7612	7619	7627
58	7634	7642	7649	7657	7664	7672	7679	7686	7694	7701
59	7709	7716	7723	7731	7738	7745	7752	7760	7767	7774
60	7782	7789	7796	7803	7810	7818	7825	7832	7839	7846
61	7853	7860	7868	7875	7882	7889	7896	7903	7910	7917
62	7924	7931	7938	7945	7952	7959	7966	7973	7980	7987
63	7993	8000	8007	8014	8021	8028	8035	8041	8048	8055
64	8062	8069	8075	8082	8089	8096	8102	8109	8116	8122
65	8129	8136	8142	8149	8156	8162	8169	8176	8182	8189
66	8195	8202	8209	8215	8222	8228	8235	8241	8248	8254
67	8261	8267	8274	8280	8287	8293	8299	8306	8312	8319
68	8325	8331	8338	8344	8351	8357	8363	8370	8376	8382
69	8388	8395	8401	8407	8414	8420	8426	8432	8439	8445
70	8451	8457	8463	8470	8476	8482	8488	8494	8500	8506
71	8513	8519	8525	8531	8537	8543	8549	8555	8561	8567
72	8573	8579	8585	8591	8597	8603	8609	8615	8621	8627
73	8633	8639	8645	8651	8657	8663	8669	8675	8681	8686
74	8692	8698	8704	8710	8716	8722	8727	8733	8739	8745
75	8751	8756	8762	8768	8774	8779	8785	8791	8797	8802
76	8808	8814	8820	8825	8831	8837	8842	8848	8854	8859
77	8865	8871	8876	8882	8887	8893	8899	8904	8910	8915
78	8921	8927	8932	8938	8943	8949	8954	8960	8965	8971
79	8976	8982	8987	8993	8998	9004	9009	9015	9020	9025
80	9031	9036	9042	9047	9053	9058	9063	9069	9074	9079
81	9085	9090	9096	9101	9106	9112	9117	9122	9128	9133
82	9138	9143	9149	9154	9159	9165	9170	9175	9180	9186
83	9191	9196	9201	9206	9212	9217	9222	9227	9232	9238
84	9243	9248	9253	9258	9263	9269	9274	9279	9284	9289
85	9294	9299	9304	9309	9315	9320	9325	9330	9335	9340
86	9345	9350	9355	9360	9365	9370	9375	9380	9385	9390
87	9395	9400	9405	9410	9415	9420	9425	9430	9435	9440
88	9445	9450	9455	9460	9465	9469	9474	9479	9484	9489
89	9494	9499	9504	9509	9513	9518	9523	9528	9533	9538
90	9542	9547	9552	9557	9562	9566	9571	9576	9581	9586
91	9590	9595	9600	9605	9609	9614	9619	9624	9628	9633
92	9638	9643	9647	9652	9657	9661	9666	9671	9675	9680
93	9685	9689	9694	9699	9703	9708	9713	9717	9722	9727
94	9731	9736	9741	9745	9750	9754	9759	9763	9768	9773
95	9777	9782	9786	9791	9795	9800	9805	9809	9814	9818
96	9823	9827	9832	9836	9841	9845	9850	9854	9859	9863
97	9868	9872	9877	9881	9886	9890	9894	9899	9903	9908
98	9912	9917	9921	9926	9930	9934	9939	9943	9948	9952
99	9956	9961	9965	9969	9974	9978	9983	9987	9991	9996
No.	0	1	2	3	4	5	6	7	8	9

Logarithms of Trigonometric Functions

38°→ ↓	sin	Diff. 1'	csc	tan	Diff. 1'	cot	sec	Diff. 1'	cos	←141° ↓
0	9.78934	16	10.21066	9.89281	26	10.10719	10.10347	10	9.89653	60
1	.78950	17	.21050	.89307	26	.10693	.10357	10	.89643	59
2	.78967	16	.21033	.89333	26	.10667	.10367	9	.89633	58
3	.78983	16	.21017	.89359	26	.10641	.10376	10	.89624	57
4	.78999	16	.21001	.89385	26	.10615	.10386	10	.89614	56
5	9.79015	16	10.20985	9.89411	26	10.10589	10.10396	10	9.89604	55
6	.79031	16	.20969	.89437	26	.10563	.10406	10	.89594	54
7	.79047	16	.20953	.89463	26	.10537	.10416	10	.89584	53
8	.79063	16	.20937	.89489	26	.10511	.10426	10	.89574	52
9	.79079	16	.20921	.89515	26	.10485	.10436	10	.89564	51
10	9.79095	16	10.20905	9.89541	26	10.10459	10.10446	10	9.89554	50
11	.79111	17	.20889	.89567	26	.10433	.10456	10	.89544	49
12	.79128	16	.20872	.89593	26	.10407	.10466	10	.89534	48
13	.79144	16	.20856	.89619	26	.10381	.10476	10	.89524	47
14	.79160	16	.20840	.89645	26	.10355	.10486	10	.89514	46
15	9.79176	16	10.20824	9.89671	26	10.10329	10.10496	9	9.89504	45
16	.79192	16	.20808	.89697	26	.10303	.10505	10	.89495	44
17	.79208	16	.20792	.89723	26	.10277	.10515	10	.89485	43
18	.79224	16	.20776	.89749	26	.10251	.10525	10	.89475	42
19	.79240	16	.20760	.89775	26	.10225	.10535	10	.89465	41
20	9.79256	16	10.20744	9.89801	26	10.10199	10.10545	10	9.89455	40
21	.79272	16	.20728	.89827	26	.10173	.10555	10	.89445	39
22	.79288	16	.20712	.89853	26	.10147	.10565	10	.89435	38
23	.79304	15	.20696	.89879	26	.10121	.10575	10	.89425	37
24	.79319	16	.20681	.89905	26	.10095	.10585	10	.89415	36
25	9.79335	16	10.20665	9.89931	26	10.10069	10.10595	10	9.89405	35
26	.79351	16	.20649	.89957	26	.10043	.10605	10	.89395	34
27	.79367	16	.20633	.89983	26	.10017	.10615	10	.89385	33
28	.79383	16	.20617	.90009	26	.09991	.10625	11	.89375	32
29	.79399	16	.20601	.90035	26	.09965	.10636	10	.89364	31
30	9.79415	16	10.20585	9.90061	25	10.10939	10.10646	10	9.89354	30
31	.79431	16	.20569	.90086	26	.09914	.10656	10	.89344	29
32	.79447	16	.20553	.90112	26	.09888	.10666	10	.89334	28
33	.79463	15	.20537	.90138	26	.09862	.10676	10	.89324	27
34	.79478	16	.20522	.90164	26	.09836	.10686	10	.89314	26
35	9.79494	16	10.20506	9.90190	26	10.09810	10.10696	10	9.89304	25
36	.79510	16	.20490	.90216	26	.09784	.10706	10	.89294	24
37	.79526	16	.20474	.90242	26	.09758	.10716	10	.89284	23
38	.79542	16	.20458	.90268	26	.09732	.10726	10	.89274	22
39	.79558	15	.20442	.90294	26	.09706	.10736	10	.89264	21
40	9.79573	16	10.20427	9.90320	26	10.09680	10.10746	10	9.89254	20
41	.79589	16	.20411	.90346	25	.09654	.10756	11	.89244	19
42	.79605	16	.20395	.90371	26	.09629	.10767	10	.89233	18
43	.79621	15	.20379	.90397	26	.09603	.10777	10	.89223	17
44	.79636	16	.20364	.90423	26	.09577	.10787	10	.89213	16
45	9.79652	16	10.20348	9.90449	26	10.09551	10.10797	10	9.89203	15
46	.79668	16	.20332	.90475	26	.09525	.10807	10	.89193	14
47	.79684	15	.20316	.90501	26	.09499	.10817	10	.89183	13
48	.79699	16	.20301	.90527	26	.09473	.10827	11	.89173	12
49	.79715	16	.20285	.90553	25	.09447	.10838	10	.89162	11
50	9.79731	15	10.20269	9.90578	26	10.09422	10.10848	10	9.89152	10
51	.79746	16	.20254	.90604	26	.09396	.10858	10	.89142	9
52	.79762	16	.20238	.90630	26	.09370	.10868	10	.89132	8
53	.79778	15	.20222	.90656	26	.09344	.10878	10	.89122	7
54	.79793	16	.20207	.90682	26	.09318	.10888	11	.89112	6
55	9.79809	16	10.20191	9.90708	26	10.09292	10.10899	10	9.89101	5
56	.79825	15	.20175	.90734	25	.09266	.10909	10	.89091	4
57	.79840	16	.20160	.90759	26	.09241	.10919	10	.89081	3
58	.79856	16	.20144	.90785	26	.09215	.10929	11	.89071	2
59	.79872	15	.20128	.90811	26	.09189	.10940	10	.89060	1
60	9.79887		10.20113	9.90837		10.09163	10.10950		9.89050	0
↑ 128°→	cos	Diff. 1'	sec	cot	Diff. 1'	tan	csc	Diff. 1'	sin	←51° ↑